Methods for Human History

Patrick Manning

Methods for Human History

Studying Social, Cultural, and Biological Evolution

Patrick Manning
World History Center
University of Pittsburgh
Pittsburgh, PA, USA

ISBN 978-3-030-53881-1   ISBN 978-3-030-53882-8 (eBook)
https://doi.org/10.1007/978-3-030-53882-8

© The Editor(s) (if applicable) and The Author(s), under exclusive license to Springer Nature Switzerland AG 2020
This work is subject to copyright. All rights are solely and exclusively licensed by the Publisher, whether the whole or part of the material is concerned, specifically the rights of translation, reprinting, reuse of illustrations, recitation, broadcasting, reproduction on microfilms or in any other physical way, and transmission or information storage and retrieval, electronic adaptation, computer software, or by similar or dissimilar methodology now known or hereafter developed.
The use of general descriptive names, registered names, trademarks, service marks, etc. in this publication does not imply, even in the absence of a specific statement, that such names are exempt from the relevant protective laws and regulations and therefore free for general use.
The publisher, the authors and the editors are safe to assume that the advice and information in this book are believed to be true and accurate at the date of publication. Neither the publisher nor the authors or the editors give a warranty, expressed or implied, with respect to the material contained herein or for any errors or omissions that may have been made. The publisher remains neutral with regard to jurisdictional claims in published maps and institutional affiliations.

Cover credit: Cindy Hopkins/Alamy Stock Photo Caption: Largo de Pelourinho, this plaza in the historic center of Salvador de Bahia, Brazil, once housed slave auctions ('pelourinho' translates as 'whipping post'). It is now a center of music and dance and a UNESCO World Heritage site

This Palgrave Macmillan imprint is published by the registered company Springer Nature Switzerland AG
The registered company address is: Gewerbestrasse 11, 6330 Cham, Switzerland

Acknowledgments

I wish to acknowledge the various experiences that have introduced me to the challenges and the pleasures of exploring academic knowledge across disciplinary boundaries. As an undergraduate at Caltech (1959–1963), I encountered ample doses of math, chemistry, and history, and smaller doses of physics, biology, English, geology, and economics, plus encouragement to link them. Working with Philip Curtin and Jan Vansina as a graduate student at the University of Wisconsin—Madison (1963–1967), I found that African History also included anthropology, linguistics, politics, and agricultural economics. I focused on economic history and took a second MS in economics with Jeffrey Williamson. Once I was teaching, William S. Griffiths introduced me to C-language programming as he guided me through my initial simulation of African population and migration. During 1978–1988, I held a Guggenheim Fellowship at the Population Studies Center at the University of Pennsylvania, where I learned from leading lights in demography.

At Northeastern University (1984–2006), I was jointly appointed in History and African-American Studies. The latter department gave me close interactions with specialists in music, theatre, visual art, sociology, political science, and economics, in a common project to study connections across the African continent and the New World diaspora. When Northeastern's graduate program in world history began (from 1994), I was able to teach a remarkable range of courses at the global level, including courses for history grads on interdisciplinary methodology in 2000 and 2001. At the University of Pittsburgh (2006–2016), I was able to teach three courses on interdisciplinary methodology, each including over 15 students from a wide range of disciplines (the syllabi are available on my website). Many of the students in those courses were outstanding. They included, in 2009, Jean Bessette, Dan Bisbee, Dan Chyutin, Racheal Forlow, Amy Hoffman, Jim Hommes, Ahmet Izmirlioglu, Liz Molnar, and Molly Nichols. Outstanding in 2012 were Sarah Bishop, John Christie-Searle, Lauren Collister, Richard Gray, Sharon

Quinsaat, Sanjana Ravi, Peter Wood, Yu Yawen, and Qi Zhang. (Sanjana Ravi was awarded a fellowship during which she co-authored a methodological article with me.) In 2015, outstanding were Matt Drwenski, Aura Jirau Arroyo, Ognjen Kohanic, Carlos Alberto López, Rongqian Ma, Jacob Pomerantz, Aisling Quigley, Bennett Sherry, and Weiyan Xiong.

In addition to these structural ties, the many individuals from whose insights and energies I have benefited include colleagues, students, and other acquaintances. At Northeastern: Yinghong Cheng, Pam Brooks, Bin Yang, Jeff Sommers, Eric Martin, David Kalivas, Tiffany Trimmer, and Deborah Smith Johnston. At Pitt: Madalina Veres, Ahmet Izmirlioglu, Chris Eirkson, Jim Hommes, and Lars Peterson. Skilled grad students coached me in genetics (Jason Carson), geology and climatology (Aubrey Hillman), and statistics (Yun Zhang, James Sharpnack, Bowen Yi, and especially Yu Liu). The faculty colleagues to whom I am most indebted are Siddharth Chandra, Jan Lucassen, Leo Lucassen, Dan Bain, Geoffrey Bowker, Molly Warsh, Ruth Mostern, Marcus Rediker, Vladimir Zadorozhny, and Hassan Karimi. I am deeply grateful for insights on the full manuscript from four readers: Chris Chase-Dunn, Eugene Anderson, Bill Wimsatt, and Felipe Fernández-Armesto. I appreciate the comments on specific sections by Ralph Adolphs, David J. Anderson, William Croft, Curtis V. Manning, David Reich, Ian Tattersall, Michael Tomasello, and Bill Wimsatt. I have been greatly pleased with the work of editor Megan Laddusaw and of Palgrave Macmillan for bringing the work to publication.

For nearly the past thirty years, my daughter, Gina Manning, has been a teacher of English. In that time, I have gained inspiration through association with her devotion to finding ways to help young people learn in a variety of circumstances—in public and private schools, for students with special needs and for those recognized as gifted. I visited classes, read her reports, and saw how many of the students made dramatic advances within each year. She sets an impressive standard; I believe I have learned from it.

Contents

1 Introduction — 1

Part I Methods for Human History

2 Human Evolution: Biological, Cultural, and Social — 23

3 Physical Science and Biological Coevolution — 33

4 Systems and Information Science — 41

5 Behavior of Individuals, Groups, and Networks — 53

6 Study of Human Institutions — 63

7 Emotions and Human Nature — 75

Part II Disciplines and Theories

8 Disciplines and Their Evolution — 83

9 Natural Selection in an Imperial Era, 1850–1945 — 91

10 DNA in a Progressive Era, 1945–1980 — 109

| 11 | Ecology and Society in a Neoliberal Era, 1980–2010 | 131 |
| 12 | Cross-Disciplinary Analysis in Global Tension, 2010–2020 | 157 |

| References | | 173 |
| Index | | 191 |

LIST OF TABLES

Table 1.1   Methods summarized, by chapter   16
Table 2.1   Assumptions in various theories of evolution   31

CHAPTER 1

Introduction

The purpose of this book is to assist students and researchers in learning and applying multiple methods for historical and cross-disciplinary analysis. It is to provide practical guidelines for expanding knowledge of the human order as we explore it in this global age. The book is to assist readers in reaching new resources—across the many fields of knowledge and back in time to moments of causation and interaction—in hopes of resolving the dilemmas of the present and the mysteries of the past. The scope of the book includes identifying major topics and issues in research, along with the disciplines, theories, methods, and data through which research questions are explored. The time frame crosses three geological epochs, including the present.[1]

This survey focuses on disciplines, theories, and especially methods of historical study. Disciplines are the social institutions and analytical engines that have divided academic knowledge into subsections, then expanding knowledge within each terrain.[2] Theories are formal interpretations of the dynamics in each field of study. Methods involve two basic stages: combining analytical logic and empirical detail to explore the dynamics of change, and then presenting the results to an audience in the hope of confirming historical and analytical interpretations.

The benefit of gaining an acquaintance with the full map of disciplines and methods associated with human history and evolution is that of expanding historical literacy. Advanced study generally provides researchers with deep

---

[1] The time frame of this work, in geological terms, includes: introduction to the Pleistocene epoch (from *Homo erectus* over 2 million years ago and including *Homo sapiens* for the past 300,000 years), the Holocene epoch (from 12,000 years ago), and the Anthropocene epoch (from 200 years ago). Much of the analysis of human evolution can be organized into these three major periods of time.

[2] By "institutions," I mean organizations composed of members who share a common set of practices and objectives.

© The Author(s) 2020
P. Manning, *Methods for Human History*,
https://doi.org/10.1007/978-3-030-53882-8_1

training in a primary specialization; in addition, advanced study commonly provides researchers with initial training in closely related disciplines. The purpose here is to expand such knowledge across disciplinary borders—to show how researchers can move beyond their principal specializations to develop broad literacy in the disciplines that address human history.

Priorities in Topics and Methods of Analysis

The discipline of history can address almost every topic in the past, but one cannot study everything at once. How are we to prioritize topics in order to explore methods for a range of historical issues that are wide yet still specific? I have chosen to give first priority to human life itself, in all its complexity, from the short term of individual lives to the long-term existence of our species. This top priority is examined in the domains of biological sciences and social sciences.

Of necessity, a secondary emphasis is the environment within which humanity exists. That is, human life is conditioned by the massive and influential surroundings, organic and inorganic, of which humanity is still a tiny part. The environment of humans includes the biosphere (the lives of animals, plants, and microorganisms) and the inorganic constituents of Earth (the geosphere, hydrosphere, and atmosphere). This second priority is examined in the domains of biological sciences (thus overlapping the biological study of humans) and also the physical and environmental sciences.

A third set of emphases comes out of intellectual accomplishments—representations of the world that have emerged from the human mind and are now centered in universities. The theories, collections of data, and interpretive arguments leave their sediment in libraries, archives, electronic files, and human brains.

These three arenas form a triad in research: human life itself, our environment, and the human knowledge on which we rely to make sense of the first two. On the third point, men and women have been asking themselves for millennia about the origins and changes in human society.[3] The discourse on origins is built deeply into myth, philosophy, religion, and culture, especially in literature. During the past 160 years, however, these questions have been rephrased, since Darwin's 1859 *Origin of Species* expressed them in terms of biological evolution.[4] The 1858 discovery of Neanderthal skeletal remains (followed by the 1864 naming of *Homo neanderthalensis*) reinforced this rephrasing of the question of human origins. Darwin's principle of natural

---

[3] David Christian, emphasizing the long human tradition of articulating myths of origin, treats evolutionary theory of life, along with the Big Bang theory of the universe, as modern myths of origin. David Christian, *Origin Story: A Big History of Everything* (New York: Little, Brown 2018), ix–x, 7–12.

[4] Charles Darwin, *The Origin of Species by Means of Natural Selection* (London: John Murray, 1859).

selection was a successful explanation of the human organism. Indeed, the logic of Darwinism became a powerful metaphor for how to organize the study of large systems in change. Parallels to Darwinian thinking became influential not only in biological subfields but in studies of the physical universe, studies of human learning, and analyses of human social change.

Within the triad of research arenas that I have identified, a common inquiry has persisted at least since the time of Darwin. I propose to think of that inquiry as a combination of biological and sociocultural evolution—I call it "the discourse on human evolution." The idea of human evolution as an encompassing and interacting set of changes, at multiple levels from the biological to the macro-social, has now moved beyond a vague dream to become a series of practical research projects. As a result, researchers and readers in many disciplines are finding that, in order to investigate the issues in human evolution, they need to develop at least basic literacy in disciplines beyond their home discipline. This broad discourse, as I present it, encompasses biological, cultural, and social evolution of humans, along with the surrounding environment—although I will emphasize that "social evolution" has involved several approaches.

During the same era of some 160 years, expanding universities have encouraged disciplinary specialization. They have led in packaging knowledge into five great intellectual containers: biological sciences, physical sciences, social sciences, arts and humanities, and now information sciences. Biological and social sciences have given most attention to aspects of human evolution, yet physical sciences and the arts and humanities have also developed visions of human evolution within their distinctive frameworks. Exploring broad evolutionary change requires crossing disciplinary boundaries—locating and learning to use resources on widely varying topics, with theories and methods that are at once distinctive and localized but also interconnected, drawing inspiration from each other. Investigating this set of issues requires attention at once to the specificity of each domain and to the interplay and generality of domains.

Disciplines have grown and expanded in depth, resulting in elaborate theories, huge libraries and datasets, impressive empirical results, specialized terminology, and institutional walls to protect each discipline from others. Yet they preserve the traces of their common origins and mutual inspirations. Each discipline is now written up skillfully by its leading practitioners. In addition, there are growing efforts at combining disciplines and showing the ways in which applying multiple disciplines can help in addressing big and difficult problems. National censuses, national income, climate analysis, genomics, and studies of health provide past cases of such large-scale, multidisciplinary enterprises. In another interesting case, the discipline of history of science formed itself in 1945, to explore historical development of the full range of natural sciences (but leaving aside the social sciences). For the future, the human order faces challenges that will require further advances in multidisciplinary analysis: the problems of impending climatic disaster, the accelerating social inequality, and, perhaps most importantly, the problems of denial in which humans have developed new knowledge but cannot agree to apply it.

To pursue this approach to methods in the study of human evolution, I must define disciplines in terms of their outward-looking relationships with each other, not just their central foci. In particular, I propose updated definitions of basic categories in evolution and its processes among humans.[5] First, I define human evolution as a general category of processes: "Human evolution is the differentiation of humans, their biological and social makeup, over multiple generations through systematic processes of internal change, in interaction with episodic changes in external environment." Specifically, I argue that human evolution involves at least three distinct mechanisms: biological, cultural, and social evolution, each with its own systemic process of internal change. In *biological evolution*, that process is the natural selection of genetic characteristics.[6] In *cultural evolution*, the systemic process is a coevolution of the process of social learning in interaction with biological natural selection.[7] In *social evolution*, the systematic process of internal change is the social selection of constructed social institutions.[8]

These restatements of the definition of evolution require a parallel restatement of the definition of coevolution. The biological definition of coevolution has recently been phrased as: "*Coevolution* is the process of reciprocal adaptation and counter-adaptation between ecologically interacting species."[9] But the perspective of human evolution requires a parallel definition for "human coevolution," including evolutionary interactions within the human species. Thus, "*Human coevolution* includes intra-species processes of reciprocal adaptation and counter-adaptation among interacting processes of evolution within the species."

---

[5] These definitions are explored in further depth in Chapter 2.

[6] The theory of natural selection is applicable to eukaryotic species that are governed by DNA.

[7] In this case, the systematic process of internal change is, inherently, the coevolution of social learning and biological evolution. The term "coevolution" was explicitly adopted by founding figures of cultural evolution theory, who defined cultural evolution as "Darwinian evolution of culture and genes." Peter J. Richerson and Robert Boyd, "The Evolution of Human Ultra-Sociality," in Irenäus Eibl-Eibisfeldt and F. Salter, eds., *Indoctrinability Ideology, and Warfare* (New York: Berghahn Books, 1998), 71–95.

[8] Details are given throughout this book and in Patrick Manning, *A History of Humanity: The Evolution of the Human System* (Cambridge: Cambridge University Press, 2020), 36–61.

[9] This definition is reworded slightly from Michael A. Brockhurst and Britt Koskelle, "Experimental Coevolution of Species Interactions," *Trends in Ecology and Evolution* 6 (2013): 367. Brockhurst and Koskelle note that coevolution "affects most organisms and is considered a key force structuring biological diversity." For an early use of the term "coevolution," see Paul R. Ehrlich and Peter H. Raven, "Butterflies and Plants: A Study in Coevolution," *Evolution* 18 (1964): 586–608.

For humans, this means that social evolution, cultural evolution, and biological evolution may interact. Coevolution in humans is restricted to adaptations and counter-adaptations among evolutionary processes. Other adaptations, especially to episodic environmental change, take place within the limits of each of the distinct evolutionary processes. Thus, the term "coevolution," as applied to humans, is not appropriate for describing the full range of interactions among the factors in history.

Further, the notion of three processes within human evolution leads logically to the definition of a Human System, encompassing the entire human species, and in which its changes are generated by the separate and interacting processes of biological, cultural, and social evolution. The Human System, including its functioning from the level of individuals and communities to its global population and structures, arises as a topic throughout this book.

Six Historical Topics, with Relevant Methods

What are the big questions in human history? What are the disciplines and theories with which we explore those questions? How did theories and methods develop and interact? The focus in this volume is on learning methods for pursuing answers to such questions.

Here are some initial examples of methods for human history—to indicate to readers what to look for in the pages to come. This section presents six topics of research and six historical methods as illustrations of the widely varying issues in analysis of human evolution. These six methods—and the additional thirty methods to be presented in Part I of this book—are presented in a structured format, beginning with a question and ending with an answer. The methods range widely in topic, theory, and details of the evidence, but I classify the methods into four types of analysis: *propose a hypothetical relationship, define an ontology, verify a hypothesis,* and *estimate parameters.* That is, they are to propose a relationship that can ultimately be tested, to categorize data to clarify their relationships, to confirm a hypothesis in comparison with an alternative, and to calculate relationships of data after assuming a verified hypothesis. Following this exploration of six historical problems, the chapter goes on to summarize the overall characteristics of historical methods, with discussion of types of data, analysis, and presentation.

The initial six topics of research range widely: from 1.6 million years ago to today's global debates, from individual tool-makers to communities creating language, and from life in the Caribbean to the endless debate about skin color. Each of these cases identifies a topic of inquiry, a historical question, a theory and discipline relevant to human evolution, a specific historical method, and a specific category of analysis. In most cases, the answer to the historical question is given in the text; in other cases, you as reader are asked to propose your own answers in what will be a "thought experiment."

Biology: change of physical appearance. Here is an important example in the biological evolution of the human form, as studied in the disciplines of paleontology and genetics. It is the study of change in human physical form from the time of *Homo habilis* to that of *Homo erectus*. Paleontology is the study of early human remains, mostly skeletal, tracing early hominin forms through changes in teeth, stature, brain capacity, male and female stature, and hunting vs. foraging activity. To understand the role of specific human remains in the process of human evolution, one must interpret those remains in terms of both genetic and epigenetic change—the evolution of DNA across generations through natural selection as compared with the *life-course development* of individuals through the actions of proteins on the genome.[10] "Turkana Boy" is the nearly complete skeleton from the edge of Lake Turkana, who died at 8 years of age, 1.6 million years ago. Compared to *Homo habilis*, his was a big shift in height, with long legs apparently adding the ability to run and hunt. Here are comparisons of Turkana Boy with averages for preceding *Homo habilis* and succeeding *Homo erectus*.

	Homo habilis	*Turkana Boy*	*Homo erectus*
Years ago	2 million years	1.6 million years	1.5 million years
Height (m)	1.35	1.6	1.8
Brain capacity	600 cc	880 cc	900 cc

What was the mechanism of change? Was it genetic evolution through a series of mutations in DNA? Or was it change in development through rapid epigenetic shifts in protein activity in a genome that changed little? This analysis involves choosing between two theories to explain the observed changes. Paleontologist Ian Tattersall argues the latter.[11] It is supported because the boy's lineage had abandoned the trees and was equipped to move only as a biped. Legs were longer, and arms were shorter than before: the boy was neither a bipedal ape nor a modern human. New lessons about the composition of DNA distinguish "coding" vs. "regulatory" DNA, where the former governs production of proteins and the latter governs the expression of coding DNA. For close relatives, differences in phenotype may owe as much to the regulation of coding genes as to the genes themselves. Changes in behavior associated with brain size, meanwhile, were to come among later communities.

---

[10] "Epigenetics" refers to changes in DNA, such as DNA methylation and histone modification, that result in heritable changes in gene expression without changing the sequence of amino acids in the DNA. The Greek prefix "epi-" means upon, near—in general, proximity.

[11] Tattersall emphasizes changing life-course development through gene expression (or regulation). Gene expression in turn changes through epigenetic changes or perhaps through a "minor mutation," as in the regulatory portion of DNA. Tattersall, *Masters of the Planet*, 98. See also François Jacob, "Evolution and Tinkering," *Science* 196 (1977), 1161–66.

> **Topic 1.1: Biological change**
> *Question:* **Did the phenotype of Turkana Boy result from genetic or epigenetic change?**
> *Theory:* genetic evolution; epigenetic life-course development.
> *Evidence:* Turkana Boy skeletal remains; materials from surroundings of skeleton; remains of other hominin individuals.
> *Analysis category:* Hypothesis test.
> *Answer to the Question:* **Turkana Boy's phenotypical changes were primarily developmental rather than genetic.** No intermediate types are known: this big change parallels other "jumps" in the fossil record. Thus, developmental change might have altered gene timing, opening "new adaptive avenues" that led to change in phenotype.
> *Presentation:* Ian Tattersall presents his confirmation of the epigenetic hypothesis in textual detail, backed up by theory, as part of a book.
> *Citation:* Ian Tattersall, *Masters of the Planet*, 91–98.

Cultural evolution: learning tool-making. Early human advances resulted from biological evolution alone. Stone tools appeared 2.6 million years ago; hands changed shape; brains expanded; stone tools are shown to have been made by right-handers. Other changes included sharpened bones; change in diet; running and throwing; and ability to move to diverse environments.

But individual learning eventually contributed: humans "crossed the Rubicon" at that point. The theory of cultural evolution argues that humans, with their large brain capacity, can learn and preserve new techniques through social learning.[12] For humans, cultural evolution is a supplement to the biological process of natural selection. Cultural evolution argues that humans could not only learn new techniques through trial and error but pass them on to the next generation, as members of the next generation learn the techniques by observation, imitation, or through mentored teaching. The new knowledge, stored in the brains of the individuals, is passed on through the process of learning rather than through genetic reproduction. Further, this theory of learning emphasizes "dual heritage," in that genetically supported cooperation among humans might grow along with social learning. The two processes of genetic growth in cooperation across the generations and social learning during individual life course do not automatically reinforce each other. But the theory assumes that their contending pressures will reach an equilibrium with the result that humans tended gradually to accumulate learning: this is "crossing the Rubicon"—it is also a description of the process of coevolution. Stephen Henning and Joseph Henrich worked, independently

---

[12] Robert Boyd and Peter J. Richerson, *The Origin and Evolution of Cultures* (New York: Oxford University Press, 2005).

and together, to test this idea.[13] They worked with modeling and game theory rather than historical data.

When did cultural evolution begin? Perhaps it was as early as 750,000 years ago (as revealed at a site of *Homo erectus* at Gesher Benot Ya'aqov, Israel); perhaps 300,000 years ago. Brains expanded to 1200 cc, and techniques for stone blades emerged. Hominids stood "on the precipice of true cumulative cultural evolution." Since the nature of the transition was necessarily quite gradual, it is difficult to pick a particular moment.

Topic 1.2: Cultural evolution
Question: **How did early individuals of genus Homo learn to make stone tools?**
Theory: Cultural evolution, assuming dual heritage, begins with "crossing the Rubicon."
Evidence: Studies of living apes; analysis of stones and bones; modeling of comparative DNA from humans and apes today and from ancient human DNA.
Analysis category: Propose a hypothetical relationship.
Answer to the Question: **Individual learning and gene-based cooperation each grew.** Human learning of techniques of tool-making was able to "cross the Rubicon," as the passing of learned techniques across the generations reinforced the genetic inheritance of practices of collaboration. This hypothesis awaits verification.
Presentation: Tables estimating the type of learning and the number of generations necessary to build a new type of learning into a population.
Citation: Shennan, *The Secret of Our Success*, 280–81, 291–93.

Evolution and environmental change: skin color. Biological evolution takes place not simply within any organism but also in interaction with the organism's environment. In an example that has become famous, British biologist Henry Bernard Davis Kettlewell conducted experiments in the 1950s showing that the rise of atmospheric coal dust in industrializing England had led to changes in certain "peppered" moths—moths of a species that was previously all-white came to be spotted with black marks, which made them less obvious to predators.[14]

---

[13] Joseph Henrich, *The Secret of Our Success: How Culture Is Driving Human Evolution, Domesticating Our Species, and Making Us Smarter* (Princeton: Princeton University Press, 2016); Stephen Shennan, *Genes, Memes and Human History: Darwinian Archaeology on Cultural Evolution* (New York: Thames and Hudson, 2003).

[14] Michael E. N. Majerus, "Industrial Melanism in the Peppered Moth, *Biston Betularia*: An Excellent Teaching Example of Darwinian Evolution in Action," *Evolution: Education and Outreach* 2 (2008): 63–67. https://doi.org/10.1007/s12052-008-0107-y. See also http://askabiologist.asu.edu/peppered-moths-game/kettlewell.

Medical science has gradually shown the difference between the internal unity of the human species and external differences in stature, hair form and color, skin color, and the size and shape of facial features. Aspects of the natural environment that can influence human phenotype include latitude and insolation, altitude, and perhaps temperature and humidity. Other environmental factors, less systematic, are diet, type and level of exercise. Explanations have not yet become definitive for differences in form of eyes, size and shape of nose and lips.

But very clearly, among humans, differences in skin color are correlated with regions of ancestry. Skin color also varies within regional populations: in recent centuries, expanded migration led to greater mixes of skin colors; the differences in skin color also led to racial categorization and hierarchy, bringing great debate, mistreatment, and confusion. Anthropologist Nina Jablonski analyzed the literature on skin color, demonstrating the specific causes of differences in skin color and the time it took for them to develop.[15] Ultraviolet radiation (UVR) from the sun causes two main dangers to humans: destruction of folate (weakening male sperm production) and destruction of Vitamin D3 (weakening calcium collection, harmful during pregnancy). Melanin expanded and created dark skins when humans lost body fur 1.2 million years ago, limiting UVR. When humans moved to higher latitudes where solar radiation was weak, decline in Melanin (lighter color) was beneficial in admitting more UVR for Vitamin D3. New data have provided an improved world map of UVR intensity.[16] The case of the Americas, settled by humans beginning 19,000 years ago, resulted in dark skins among those at the equator. This suggests that the development of melanin among light-skinned people took place well within 15,000 years.

---

[15] Nina Jablonski, "The Evolution of Human Skin and Skin Color," *Annual Review of Anthropology* 33 (2004): 585–623. See also Nina G. Jablonski and P. G. Maré, eds., *The Effects of Race* (Stellenbosch: African Sun Media, 2018).

[16] For a world map showing the varying degrees of insolation that affect skin color, see "Evolution of Human Skin and Skin Pigmentation," Anth.la.psu.edu/research/research-labs/jablonski-lab. As the accompanying text notes, "Our research on the evolution of human skin and skin color has demonstrated that skin color is the product of natural selection acting to regulate levels of melanin pigment in the skin relative to levels of ultraviolet radiation (UVR) in the environment. Melanin is a natural sunscreen that prevents the breakdown of certain essential biomolecules (in particular, the B vitamin folate, and DNA), while permitting enough UVR to enter the skin to promote the production of essential vitamin D."

> **Topic 1.3: Evolution and environmental change**
> *Question:* **What are the causes of human skin color and its variations?**
> *Theory:* Genetic evolution.
> *Evidence:* Skin colors, UVR levels, folate levels, production of Vitamin D.
> *Analysis category:* Estimate parameters (in rank order)
> *Answer to the Question:* **Skin color was determined by natural selection in response to different environments.** Melanin in skin protects against UVR, causes dark skin, and has health benefits.
> *Presentation:* Historical summary of human loss of fur, darkening of skin, then lightening and darkening of skin as humans moved to higher and lower latitudes, showing implications for health.
> *Citation:* Jablonski, "The Evolution of Human Skin and Skin Color," 585–86, 599–602.

Social institutions: language. The ability to speak complete sentences depends on skills in reasoning, an interest in communication, and ability to vocalize those two clearly. Of these, speaking in sentences was the last step and is argued to have begun 70,000 years ago.

The theory of institutional evolution argues that the formation of the first speech community also created the first social institution, where an institution is a group of people cooperating to achieve a common purpose. This exercise asks you to imagine yourself in a group of humans that are trying to develop a system of spoken language. It asks you to (1) assume that people lived in families of 10–20, without spoken language, and (2) then assume that young people (ages 10–15) began to speak in sentences, developing their spoken communication with each other. What steps would members of the new group take to ensure that their group survives and expands? What resulting changes might take place in the pre-existing family structure?

The citation from Manning gives ideas on some of the early steps. Was there social conflict as the new group formed? How was speech taught to adults, children, and infants? How big did the group need to be? How did speakers select words and build sentences for names, body parts, roles; actions of work and play; modifiers—big and small, old and young? Speakers needed to stay in touch, accepting additions and changes to the language. Creation of a speaking community was a rehearsal for the process of creating groups for other purposes. Since families had taken form much earlier under the influence of genetic evolution, any changes in family structure resulting from the rise of language would have been through a process of coevolution.

> **Topic 1.4: Social institutions and coevolution**
> *Question:* **How did spoken language create an expanded community?**
> *Theory:* institutional evolution, assuming that groups create and renew social institutions; biological evolution of families
> *Evidence:* Your experience on learning and speaking language, and on social relationships among speakers of a language.
> *Analysis category:* Propose hypothetical relationships.
> *Answer to the Question:* (Thought experiment) **List your ideas on decisions made by individuals and groups of people beginning to speak.** Explain which decisions could create a larger community that sustains language. Identify any expected change in families.
> *Presentation:* Present an orderly list, reflecting your understanding of decisions made within a group that would preserve a language community.
> *Citation:* Manning, *A History of Humanity*, 37–43.

Large-scale datasets: social inequality. Discussion of social inequality, focusing mainly on recent lifetimes, has developed some large datasets. A leading example, the *World Inequality Database*, led by Thomas Piketty, focuses on national-level data from the twentieth and twenty-first centuries.[17] Thanks to the United Nations and the World Bank, world population and economy are well documented since 1950. Before that time, only Europe and North America are well documented, plus Japan and China. For now, it is hard to know the experience of inequality in other places and times.

Here is a proposal for research on an area of the world that was at once a center of wealth and poverty: the islands of the Caribbean Sea, from 1500 and especially 1750 to the present, from the era of plantation agriculture to the present-day mix of independent nations and colonies. Here is a way to explore the population, economy, and society of a region, outside Europe, that experienced wide-ranging and changing inequality and is thoroughly documented with European records.[18]

Caribbean territories were clearly marked by island shores, islands were occupied by six different powers and numerous independent nations, and documents exist for most times and spaces on commerce, wars, administration, population, migration, rates of birth and death, disease, climate, and even land use. (The data need to be retrieved, translated, converted to consistent measures, and organized to reflect the different sorts of

---

[17] World Inequality Database, http://WID.world.

[18] Matt Drwenski, "Scales of Inequality: Strategies for Researching Global Disparities from 1750 to the Present" (MA thesis, University of Pittsburgh, 2015).

economic, social, health, and political inequality.) From 1500, Taino and other indigenous farming peoples met conquest by Spain and then adventurers of English, French, Dutch, and Danish origin. Empires captured and exchanged islands. Sugar dominated agriculture from the late seventeenth to mid-nineteenth centuries, followed by coffee, tobacco, cotton, cattle, and banana production. Immigration came from Africa by slave trade, from India by indenture, and from Europe. Wealthy landowners built fortunes, some living in the Caribbean and others as absentee landowners. Haiti fell from a rich to a poor territory. Mineral extraction grew—petroleum, iron, and aluminum ores. From 1940, migrants left the Caribbean steadily.

> Topic 1.5: Large-scale datasets
> *Question:* **How serious has inequality been in Caribbean history since 1750?**
> *Theory:* Economic and social history
> *Evidence:* Data on Caribbean territories on population, migration, production, commerce, government, climate, and disease since 1750. Data must be classified into an ontology.
> *Analysis category:* Propose a hypothetical relationship
> *Answer to the Question:* **Inspection of data shows concentrations of wealth and poverty, with significant change over time.** Combination of Caribbean data should reveal new dynamics.
> *Presentation*: This will be a proposal for a research project to collect, process, and analyze data.
> *Citation:* Caribbean Data Portal, http://www.caribbeaneconomics.org/about; Piketty, *The World Inequality Database*, wid.world[19]

Human System: gender relations. The Human System is defined here as the aggregate of all humans and their interactions over time and space.[20] It collects raw materials, nurtures humans with food, clothing, and culture, and expels human waste and social waste. It functions through subsystems that interact to process information, process matter and energy, and reproduce the system. Institutions, guided by their members, direct all of the subsystems and functions within the system overall. Yet the system encounters difficulties because of internal inefficiency or breakdown in subsystems, or because of changes in the outside environment.

Recent research has revealed a major inefficiency in the Human System. On the one hand, the basic equality of human females and males has been

---

[19] Within wid.world, see wid.world > Methodology > Methodology; see also wid.world > About Us > WID.WORLD.

[20] On elements of the Human System, see Chapter 4; see also Manning, *A History of Humanity*, 52–61.

demonstrated: the equality of females in intelligence, skill, and productivity, especially as they also bear the children that preserve the human species. On the other hand, females undergo discrimination and neglect in many areas. Human welfare and knowledge would advance if women were able to participate as equals. Despite this realization, discrimination against females persists.

What do you think are factors within the Human System that cause gender discrimination to continue despite this knowledge? How can one break gender relations into their main elements, identify relevant theories and variables, identify methods for analysis, and sketch application of the method? Many people argue that discrimination arises out of the natural order; they reject any causal human involvement in the phenomena. What would be a research design to explore this general phenomenon? Can women attend school? Can females achieve equal earnings? What about work of family care, home care? How to allow for realities of childbirth and child care, sexuality and sexual desire? The Human System is supposed to function well—how does it fall into functioning poorly? We now understand the costs of gender discrimination—too much work, not enough pay, personal exploitation (physical and sexual), prejudicial treatment. In what ways does the system support these? Does the system try to reduce exploitation? Build a hypothesis that could be tested.

Human gender relations can be seen to have been structured by each of the processes of biological, cultural, and social evolution, by their coevolution, and by episodic environmental factors. This framework provides a distinctive way to consider the issue of gender relations. The question asks you to think especially about the effects of evolving social institutions on gender relations, but your response can include ideas on the influence of biological and cultural evolution as well as environmental change.

Topic 1.6: Human System behavior
Question: **What systemic processes sustain gender discrimination today?**
Theory and assumptions: Institutional evolution (assuming that institutions have processes to correct their errors and inefficiencies)
Evidence: Observation of current social institutions and their operation
Analysis category: Propose hypothetical relationships.
Answer to the Question: (Thought experiment) **Provide a list of functions, organizations, or beliefs within the Human System that reinforce gender discrimination.**
Presentation: A description of patterns of gender discrimination as affected by social structures, but perhaps including biological and cultural factors as well as environmental changes.
Citation: "Gender Equality: United Nations," www.un.org/sections/issues-depth/gender-equality.

Some Characteristics of Historical Methods

The above examples point to the range of historical methods, topics, and scales of analysis. One may then ask what such wide-ranging historical methods have in common.

To begin, I identify three broad *elements of historical reality* to which historical methods apply—elements that researchers seek to elucidate with theory and method. Initially, there is the *past*, the events and processes of the past as they actually unfolded. The past is over and will not change. Nevertheless, our view of the past changes repeatedly—changing as we bring new evidence to bear and view the past through additional lenses. Second, there is the *available evidence* on the past: such evidence is necessarily partial but it can be voluminous. Evidence can also be hypothetical. Mostly, we find that evidence on the past grows as we build our archives, but available evidence also declines, for instance as documents are lost or well-informed individuals die. Third is the *lens of theory and hypothesis* through which we view the past. Theories and interpretive frameworks change in response to adjustments in their own internal logic but also in response to empirical studies and shifts in historical philosophy.

Each historical method, in its own way, reflects the collection of evidence about the past. The method is a process for analyzing and scrutinizing evidence through the lens of theory and hypothesis in order to construct a representation of the actual past. The discussions in this book are intended to provide information and an invitation to study further the range of available methods. Here are eight points and some sub-points that add up to the steps in a historical method.

1. Topic and discipline of study. The researcher selects these.
2. Question for analysis. The researcher selects this question or finds that it has been suggested by others.
3. Theory to be applied. The theory gives the logic of analysis but also the scope of the study in time, space, and scale of aggregation. The theory's logic of analysis includes its assumptions and its dynamics of change.
4. Evidence. The data—information on both past and present (including past analyses)—must be defined, collected, and edited to achieve their clarity and consistency.
    a. Collect data. These are data relevant to the scope and theory of study, relying on the practices of the discipline.
    b. Categorize data. The evidence may be qualitative or quantitative, held in structured or unstructured form. In quantitative terms, data may be classified as nominal, ordinal, or interval. Nominal (or categorical) data are descriptive and unique; ordinal data may be organized in order, as with the order of siblings; and interval data may be given precise numerical values, as in the size of a population.

c. Ontology. The data are organized and sorted into an ontology, a classification of the data—this is where the conceptualization and the data encounter each other.[21]
   d. Missing data. These are cases of evidence that is important to the analysis but not available in the historical record. In response to these absences, a common tactic is to estimate or simulate the missing data, perhaps relying on theory to estimate the missing data (thus introducing a certain circularity into the analysis).
   e. Proxy data. The data available for past phenomena may not fit logically with the variables identified in the conceptualization, so that the researcher must treat the available data as a "proxy" for the variable under study, making an argument for the closeness of fit of proxy and target variable.
5. Analysis category. Select among four main types of answers to the questions posed in historical analyses:
   a. *Propose a hypothetical relationship.* The analyst explores relevant data in order to propose a relationship—that is, a hypothesis on the structure or dynamic linking data within the evidence.
   b. *Define an ontology.* The evidence is sorted into categories that reveal relationships emphasizing similarities among items within a category, and contrasts among items in different categories.
   c. *Verify a hypothesis.* A hypothetical answer is posed to the question, within the conditions of the theory. Evidence is organized to show whether the hypothesis or another selected possibility is most likely. In statistical analysis, hypothesis-testing calculates a level of confidence in the answer.
   d. *Estimate parameters.* A hypothesized relationship is assumed, and analysis estimates the strength of the relationship in the cases included in the evidence.
6. Answer the question posed. That is, analyze the data using the theory and type of analysis selected. The answer should be straightforward but should include a full explanation of how it was developed and why it is the best response.
7. Presentation. The form of the presentation must fit with the theory and the type of analysis. The presentation may be descriptive, placing data into the ontology for observation. Or it may be analytical, in effort to determine a relationship among factors or variables within the data under study. In the latter case, the data are subjected to a chosen technique, perhaps with the formal test of a relationship among variables, so that the results verify or reject the proposed relationship.
8. Citation. Reference to resources used in applying this method, so that other researchers may replicate the results.

[21] For conceptualization of historical data, see Patrick Manning, "Epistemology," in Jerry H. Bentley, ed., *Oxford History Handbook: World History* (New York: Oxford University Press, 2011), 105–21.

Table 1.1 presents a guide to the methodological summaries in this book. It lists the 36 methods that are summarized in the Introduction and Part I of this book, by chapter, with the label for each method, the question posed, and brief statements of the hypothesis and conclusion. Further, in research

Table 1.1 Methods summarized, by chapter

	Topic	Question	Theory	Analysis Category
Chapter 1. Introduction				
1.1	biological change	Did the phenotype of Turkana Boy result from genetic or epigenetic change?	genetic evolution; epigenetic life-course development	hypothesis
1.2	cultural evolution	How did early individuals of genus Homo learn to make stone tools?	cultural evolution	relationship
1.3	evolution and environmental change	What are the causes of human skin color?	genetic evolution	parameters
1.4	social institutions	How did spoken language create an expanded community?	institutional evolution	relationship
1.5	large-scale datasets	How serious has inequality been in Caribbean history?	economic & social history	relationship
1.6	Human System behavior	What systemic processes sustain gender discrimination today?	institutional evolution	relationship
Chapter 2. Biological - Cultural - Social				
2.1	biological evolution	What information does whole-genome analysis give on human ancestry?	genetics	parameters
2.2	cultural evolution	What is the balance of genetics and learning in the dual heritage of cultural evolution?	cultural evolution	parameters
2.3	social institutions	What institutions were created to enable agriculture to become successful?	institutional evolution	relationship
Chapter 3. Physical sciences				
3.1	physics	How do we know the age of the Earth?	physics, chemistry, geology	parameters
3.2	chemistry, archaeology	How does radiocarbon analysis determine the date of human settlement in South America?	chemistry, radiocarbon analysis	parameters
3.3	Earth sciences	How do ice cores document weather 10,000 years ago?.	Earth science; chemistry	parameters
3.4	environmental coevolution	What are two examples of successful relationships between a type of pollinator and a type of flower?	ecology	relationship
Chapter 4. Systems & Information Sciences				
4.1	general systems	How do ingest and output stay in balance in open systems?	systems	relationship
4.2	Gaia	What conflicting forces brought temperature equilibrium in Gaia	Gaia theory, Earth system	relationship
4.3	living systems	How do subsystems keep a human organism alive?	living systems; biology	relationship
4.4	Human System	What problems in subsystems might cause the Human System to collapse?	institutional evolution; living systems;	relationship
4.5	World-Systems	What does world-systems theory explain about Akkadian conquest of Sumeria?	World-System theory	hypothesis
4.6	archives	how can one locate topical resources within PubMed?	library science	ontology
4.7	ontology	What should be the structure of a geographical ontology for your home town?	ontology	ontology
4.8	databases	Do human societies from around the world exhibit similarities in the way that they are structured?	"theory neutral"	hypothesis
Chapter 5. Behavior				
5.1	individual behavior	What is the best strategy for playing Prisoner's Dilemma?	game theory	hypothesis
5.2	group behavior	To create a small private school for young children, show how would you identify the members, goals, and beneficiaries of the new institution?	collective intentionality	relationship
5.3	interactions - groups & individuals	Based on your experience, what criteria can show that several people are acting as a group?	collective intentionality	relationship

Table 1.1 (continued)

5.4	networks of individuals	What are differences among ego-networks, dyads, and triads in network interactions?	network theory	ontology
5.5	networks of groups	What are differences in the organization of groups within Catholic, Sunni Islamic, and Pentecostal faiths?	network theory; collective intentionality	hypothesis
Chapter 6. Social Institutions				
6.1	linguistics	How long does it take for words to be borrowed across languages?	historical linguistics	parameters
6.2	migration	What institutions are highlighted in migration theory?	migration theory	ontology
6.3	exchange & commerce	In the transatlantic commercial system of the eighteenth century, identify aspects of it that were institutions and aspects that were networks.	economics; sociology; history	ontology
6.4	institutions of leadership	What are the relative strengths of achievement-based and hierarchy-based leadership?	social anthropology	relationship
6.5	literacy	How did literacy spread?	institutional evolution	relationship
6.6	capitalism	What new institutions enabled capitalism to expand in the twentieth century?	institutional evolution	relationship
Chapter 7. Emotions				
7.1	emotions in individuals	How are feelings related to emotions?	emotions in biological context	relationship
7.2	emotions in groups	How do humans construct their emotions?	constructivist psychology	relationship
7.3	human nature in biological terms	Does there exist a biologically based human nature?	biology, psychology	hypothesis
7.4	human nature in social terms	Is there a human nature specific to group behavior?	constructivist psychology	hypothesis

addressing human behavior and its history, it is wise to consider placing humans in the context of other animals. Humans are distinctive, but not as unique as we sometimes like to think. Cases of human-animal interaction, especially in the case of domesticated animals, show how much can be shared between humans and animals. Which human behaviors are common to humans but otherwise unique? Which behaviors are shared with other animal species? Which human behaviors are social rather than biological in origin, so that they can be changed through consciousness and nurture? In migratory behavior, birds, fish, and mammals have various sorts of group behavior that were constructed without the aid of speech. For emotions, the human ability to express feelings in speech doubtless transforms and amplifies emotions, yet the parallel behavior of animals in varying situations indicates that there is a biological underpinning to human emotions.[22]

Organization of This Book

The six chapters in Part I provide a cross-sectional survey of present-day methods in historical studies, setting roughly 30 specific methodologies in the context of broader disciplines and theories. The five chapters in Part II trace two centuries of the development and interactions of research traditions that can now be seen as converging toward seeking a coherent interpretation of human evolution.

---

[22] More broadly, work with historical data requires a balance of attention to theory, ontology, raw data, processed data, analysis, and verification. Patrick Manning, "Analyzing World History," in Manning, ed., *Navigating World History: Historians Create a Global Past* (New York: Palgrave Macmillan, 2003), 313–23.

Part I is a concise summary of major methodological domains, with references for further study. It focuses on human evolution and living creatures more broadly, for each of several methodological arenas. Each chapter is divided into several sections, addressing different disciplines within a common category. In laying the groundwork for each methodology, the disciplinary sections identify the scope of study, give definitions of main terms, and provide summaries of the theories and principles with which they are associated, types of analysis, and citations of more detailed methodological works for further reading. Biological sciences are the focus of Chapter 2, including a survey of three types of evolutionary process. Physical sciences and environmental studies are surveyed in Chapter 3. Chapter 4 begins with general systems and social systems, and then turns to information sciences. Chapter 5 compares human behavior as individuals, in groups, and in networks—all are essential issues for the analysis of social evolution. Chapter 6 presents social science disciplines, centering on the analysis of institutional evolution.[23] The chapter then turns to issues in the disciplines of anthropology, economics, history, linguistics, politics, and sociology. Psychology is discussed in Chapters 5 and 7. Disciplines of the arts and humanities are discussed briefly in Chapters 2 and 6.

Part II is a chronological review, from 1850 to the present, of the developments in the various disciplines. Its chapters point out underlying philosophical and analytical principles and how they have influenced each other. Chapters 8 through 12 include sections on specific fields, in alternation with sections tracing the interactions among disciplines. These chapters, while centered on each period's thinking within academic disciplines, also review the overall transformation of the human order in the same periods—in social change, public opinion on social priorities, and ideological disputes. For instance, because the nineteenth and twentieth centuries were an era of European domination of the world and its academic institutions, there is a Eurocentric flavor and a male-dominated aspect to the interpretations. Perhaps this bias can be limited through continued historical analysis. For instance, long-term interpretations of changes in ideas are already showing that knowledge has been exchanged among all regions and segments of human society, rather than arising only among elites in centers of excellence.[24]

Further elements of this book provide additional materials on disciplines, theories, and especially methods. In addition to notes and a list of works cited, a list of methodological reference works and websites gives examples

---

[23] Methods presented in Chapter 6 address the issues of language, migration, exchange and commerce, government, literacy, and capitalism.

[24] Patrick Manning and Abigail Owen, eds., *Knowledge in Translation: Global Patterns of Scientific Exchange, 1000–1800 CE* (Pittsburgh: University of Pittsburgh Press, 2018).

of details in several disciplines. Two outstanding examples are *Deep History*, edited by Andrew Shryock and Daniel Lord Smail, which provides essays on methods for long-term human history, and *Mathematical Models of Social Evolution*, in which Richard McElreath and Robert Boyd survey the models used in cultural evolution.[25] In a fortuitous development, introductory texts on the methods of the various disciplines have now become clearer and more widely available. Because scholars in many fields have sought to solicit funding and new students, they have put energy and imagination into writing texts that provide coherent review of major methods and summaries of recent research results. The leading academic journals are changing their formats, now requiring statements of methodologies, details on data to back up the analysis, and clear tables and graphs. Science journalists provide basic summaries and guides to current research; literature reviews occasionally summarize decades of advances. The theories remain as difficult as ever, but it has nevertheless become much easier than before to read in fields outside one's own specialization.

---

[25] Andrew Shryock and Daniel Lord Smail, eds., *Deep History: The Architecture of Past and Present* (Berkeley: University of California Press, 2011); Richard McElreath and Robert Boyd, *Mathematical Models of Social Evolution: A Guide for the Perplexed* (Chicago: University of Chicago Press, 2007). Topics addressed in *Deep History* include energy, kinship, language, food, migration, and scale.

PART I

Methods for Human History

CHAPTER 2

Human Evolution: Biological, Cultural, and Social

Based on recent research and rethinking, it can be argued that humans undergo evolution at three different scales in their existence. The three scales are biological evolution (based on the workings of DNA), cultural evolution (relying on learning practices stored in individual brains), and social evolution (based on the workings of groups in collective intentionality).[1] Although the three processes are quite different in detail, Darwin's notion of "descent with modification" can be seen to characterize them all. Each of the three processes, within its own domain, has led to remarkable differentiation. Further, the combination of the three processes has led to accelerating change and to the expansion of a Human System that incorporates all three mechanisms of change.

It is important, nevertheless, not to leap into labeling such evolutionary processes as "progress," as Herbert Spencer tried to do in the era of Darwin. It is largely true that the individual changes at each of the three levels served formally to increase the degree of "fitness" at that level. Yet it is quite a different matter to argue that the sum total of the changes was an "improvement" for humanity or for life in general. The concern in this volume is the specific mechanisms of change over time—at biological, cultural, and social levels. Quite a different issue is the aggregation of individual changes into an understanding of overall change for humanity and the Earth. That task, important in history overall, cannot be handled successfully by skipping over the details at each scale in the processes of change. Therefore, the question of "progress" for humanity as a whole must be separated from the specifics of biological, cultural, or social evolution. The point here is that the issue of progress is a substantially different question from the process of "descent

---

[1] Collective intentionality and institutional evolution comprise a particular approach to social evolution. Chapters 10 through 12 set this approach in the context of other visions of social evolution.

with modification," leading to increasing diversity in forms, which is explored here in its biological, cultural, and social dimensions. The issue of progress—of directional change at the macro-level—is important but distinctive.

This chapter begins with the logic of biological evolution and then shows ways of extending Darwinian logic to the related issues of cultural and social evolution. As Darwin phrased it, the core of his analysis relied on variation in biological characteristics, selection of characteristics through a struggle for existence, and inheritance of characteristics—as these factors are judged by the standard of fitness that is the degree of reproduction of individuals and the species. Darwin's initial phenotypical analysis of individual organisms has now been extended to other scales of life, giving us genetic and molecular details that provide deep new insights yet do more to reaffirm than to revise Darwin's original insights. In recent years, analysts are turning to exploring questions that Darwin raised in his later books on the descent of humanity and on emotions in animals and humans. It remains plausible that Darwin's founding principles are valuable as well at these additional levels.

Next, I explore cultural evolution, a field of study that has grown rapidly since 1980, in which analysts use analogies to Darwinian principles to explore processes of social learning that are archived in human brains, paired with the evolution of population characteristics in the genome. This type of analysis, applied especially to *Homo sapiens* in the time before speech arose but also late Pleistocene and Holocene times in which humans were able to speak, focuses on documenting the gradual pairing of genetically supported cooperative tendencies with social learning that facilitates improvement in tool-making and other technologies.

For social evolution, I argue that new formulations in the study of group behavior gain benefits from analogies to Darwinian reasoning. In this approach, social institutions are seen as creations by groups of humans under the logic of collective intentionality. These institutions serve as a social parallel to the DNA of biological evolution. That is, institutions generate the activities that preserve society, they vary because of the innovative skills of individuals and groups, they are selected for survival based on their ability to provide advantages for their beneficiaries, and institutional characteristics are inherited based on their ability to archive and reproduce their practices.

At the conclusion of this chapter, I present two sorts of comparison of these three frameworks for evolutionary thinking. First, I summarize the assumptions made in each of the three frameworks, showing their underlying parallels but also their specific differences. Second, I describe the different definitions of "culture" and different approaches to culture in each of the three frameworks.

BIOLOGY AND BIOLOGICAL EVOLUTION

The term "biological evolution" applies to all eukaryotic organisms—initially single-celled organisms containing a nucleus (with chromosomes, genes, and DNA) and organelles. The ancestors of today's plants, fungi, and animals separated from each other roughly 1.5 billion years ago; multicellular life began roughly 900 million years ago.[2] Biological evolution centers on the genetic change in species through natural selection but has been extended to include ontogenic change, the life-course development of species (discussed in Chapter 1 for the case of Turkana Boy).[3]

In Darwinian theory of biological evolution, the governing principle is "descent with modification" through the process of natural selection. The principles of natural selection are traced within the individual organism, which is the *unit of evolution*. Those principles are enunciated in terms of variation, selection, inheritance, and fitness. *Variation* takes place most fundamentally in the DNA of the organism, where the nature of genetic variation is primarily random change. But variation also takes place through epigenetic (or ontogenic) change; the dynamics of epigenetic change are still being worked out.[4] *Selection* among biological variations takes place through natural selection, a term that encompasses a wide range of potential changes, from rejection of a mutant form within a cell to rejection of otherwise successful variants that led to few offspring. *Inheritance* takes place through DNA, as the genome (in association with epigenetics) both retains any successful variations and reproduces them through mitosis. *Fitness* of any variation is assessed by the level of biological reproduction of organisms carrying the variation.[5]

Sequencing equipment for determining the structure of DNA has become more and more powerful, enabling widespread analysis of human and other genomes.[6] Nonetheless, there remain problems in the details of genetic analysis of the human past. In particular, when DNA samples are taken from living humans, one must classify the present-day sample source, projecting that genome to classifications of past populations. Both classification and projection require consultation with social scientists. Analysis of *ancient DNA* clarifies but does not resolve this problem: the location of the ancient genome is known, but that may not be the same place where that person's early

---

[2] Michael Marshall, "Timeline: The Evolution of Life," *New Scientist*, 14 July 2009. See also Peter J. Bowler, *Evolution* (Oakland: CA: University of California Press, 2009): first published in 1984 as *Evolution: The History of an Idea*, this is effectively the sixth edition of this important review, which is strongest in its analysis of the period up to 1980.

[3] Darwin, *Origin of Species*; Stephen Jay Gould, *Ontogeny and Phylogeny* (Cambridge, MA: Belknap Press of Harvard University Press, 1977); François Jacob, "Evolution and Tinkering," *Science* 196 (1977): 1161–66. Infectious disease arises from biological evolution of microorganisms (viruses, bacteria, fungi, and parasites) that occupy plants and animals as hosts. Jared Diamond, *Guns, Germs and Steel: The Fates of Human Societies* (New York: Norton, 1997).

[4] Gould, *Ontogeny and Phylogeny*.

[5] For examples of continuing biological evolution of humans, see Edmund Russell, *Evolutionary History: Uniting History and Biology to Understand Life on Earth* (New York: Cambridge University Press, 2011).

[6] The instruments used in laboratory analysis of biological evolution include electron microscopes; analysis of nuclear magnetic resonance has become central to displaying the conformation of proteins.

ancestors lived.[7] A second problem in genetic analysis of the past is that of the molecular clock. That is, genomic analysis gives precise ordinal results, confirming the order in which genetic divergence has occurred, but does not give the specific chronologies required for comparison to other data.

Whole-genome analysis is analysis of locations throughout the human genome, in contrast to studies of specific locations such as Y-chromosomes (which are particularly valuable for studying male history) and mitochondrial DNA (for females). Svante Pääbo, Swedish born and working in Germany, led a group that developed many of the experimental techniques for whole-genome ancient DNA analysis, and applied them for the most part to the study of archaic humans like Neanderthals and early modern humans. Eske Willerslev, in Copenhagen, and David Reich, working at Harvard, have focused on applying these techniques to large numbers of individuals living within the last ten thousand years. Using the techniques that David Reich has focused on, the efficiency is such that the cost for analysis of an individual genome has fallen to less than 500 US dollars.

Reich, in a book summarizing much whole-genome research, emphasizes new results showing early and unexpected divisions in the human ancestry. These include discovery of the Denisovans, ancestors of Neanderthals, more distant "ghost" populations, and newly discovered lineages of population in Africa.[8] Asia is shown to have had a greater role in human evolution than previously thought, yet Africa appears as the principal region of human emergence. A second major emphasis is in genetic mixing across populations, in both early and recent times. Analysis of the more recent Holocene period reveals migrations affecting Africa, the Americas, and showing parallels of India and Europe.

Topic 2.1: Biological evolution
Question: **What information does whole-genome analysis give on human ancestry?**
Theory: genetics (including technology of sequencing and molecular clocks)
Evidence: Details of genome; time and place of sample.
Analysis category: Estimate parameters.
Answer to the Question: **Whole-genome analysis includes composite pictures of the ancestry of each genome, both long term and short term.** Refer to parameters
Presentation: Presentation involves setting priorities between long-term and short-term analysis.
Citation: Reich, 1–6, 10–14, 53–59.

---

[7] It may be that pairing modern and ancient DNA studies will help in identifying the birthplace of the ancient person whose DNA is sampled and the location of that person's earlier ancestors of the study. David Reich, *Who We Are and How We Got Here* (New York: Pantheon, 2018).

[8] Reich, *Who We Are*, loc. 1607–1840.

ECOLOGY AND CULTURAL EVOLUTION FOR GENUS *HOMO*

The field of ecology expanded rapidly in the late twentieth century through multidisciplinary analysis linking species of plants and animals and the physical world.[9] Out of that ecological framework came the new discipline of cultural evolution. The theory of cultural evolution focuses on social learning, where it is known as "dual heritage," in that genetic and cultural evolutionary processes coevolve. The term "cultural evolution" was developed for study of humans, though it may be applied to primates and other animal species. In the study of learning, accretions to learning are considered to constitute cultural elements. As Boyd and Richerson defined the term in their first book, "By 'culture' we mean the transmission from one generation to the next, via teaching and imitation, of knowledge, values, and other factors that influence behavior."[10]

One may apply the original Darwinian evolutionary terms to cultural evolution, with slightly different meanings. In cultural evolution, the *unit of evolution* is again the individual organism, now including both biological evolution in the genome and cultural evolution in the brain.[11] The *variation* takes place both in the genome and in the brain of the organism; the source of variation is random change in the genome but also individual choice in learning by human organisms, including reciprocal behavior. *Selection* of surviving innovations arising from social learning can take place according to three processes, in which genetic selection is paired with: (i) personal selection through individual learning; (ii) inclusive selection on learning or reciprocal behavior; and (iii) multilevel selection on learning and reciprocal behavior.[12] The *inheritance* of innovations is archived and retained in the genome and the brain of the individual; the dual inheritance is conveyed genetically through DNA and culturally by exchange from brain to brain across the generations through learning practices. *Fitness* may be determined at three levels: (i) the personal rate of reproduction of organisms carrying the cultural and biological variation; (ii) the rate of reproduction of organisms included in the behavior of an individual organism through inclusive fitness; and (iii) the rate of reproduction of a group of organisms through the above processes. These three scales of cultural analysis are argued to be equivalent to each other. There are effectively four subfields of analysis within the study of cultural evolution. First has been the theorization of the dual-heritage

---

[9] Within ecology, the field of ethology focuses on animal behavior, both in laboratory and in field studies.

[10] Robert Boyd and Peter J. Richerson, *Culture and Evolutionary Process* (Chicago: University of Chicago Press, 1985), 2.

[11] Boyd and Richerson, *Culture and Evolutionary Process*.

[12] On inclusive selection, see Chapters 5, 10, and 11; also McElreath and Boyd, *Mathematical Models*.

process, relying especially on game theory and population dynamics. Second has been the experimental analysis of human tool-making in mid-Pleistocene hominin populations. Third is a set of studies of "ultrasociality," focusing on *Homo sapiens* in Holocene and even Anthropocene times.[13] Fourth is the range of other approaches indirectly linked to cultural evolution, especially memetics, studies of visual communication, and studies of primate life-course development.[14]

These acts of learning, whether by trial and error or by imitation, are related to human imagination and ideas.[15] They are directly relevant to the example of making stone tools in Chapter 1.

Topic 2.2: Cultural evolution
Question: **What is the balance of genetics and learning in the dual heritage of cultural evolution?**
Theory: dual-heritage theory in cultural evolution.
Evidence: models of practice of learning and its interaction with genetic composition.
Analysis category: estimate parameters
Answer to the Question: **Since "crossing the Rubicon," learning practices and culture have become predominant.** They have led in setting the directions of human genetic evolution.
Presentation: modeling the results.
Citation: Boyd and Richerson, *Culture and Evolutionary Process*, 3–11, 172–78.

Social Sciences and Institutional Evolution

The term "evolution" has been applied to social change by many authors. L. H. Morgan advanced a stage theory of social evolution that gained wide attention; Leslie White revised and expanded Morgan's approach, focusing on the transition to agriculture. According to the various approaches, social

---

[13] Boyd and Richerson, *Culture and Evolutionary Process*; McElreath and Boyd, *Mathematical Models*; Peter Turchin, *Ultra Society: How 10,000 Years of War Made Humans the Greatest Cooperators on Earth* (Chaplin, CT: Beresta Books, 2015).

[14] Nicole Creanza, Oren Kolodny, and Marcus W. Feldman, "Cultural Evolutionary Theory: How Culture Evolves and Why It Matters," *PNAS* 114, 30 (2017): 7782–89. In addition, emotions arise as expressions of individuals and may play a role in cultural evolution. For an introductory discussion, see Chapter 7.

[15] See Turchin, *Ultra Society*. This view of cultural evolution places a premium on uniformity within communities and variety among communities, to facilitate selection of the fittest communities; hence, migration brings dangers to uniformity of communities. For further details, see page 163.

evolution may have begun as early as 70,000 years ago with the rise of language; 10,000 years ago with the rise of agriculture; or 5000 years ago with the rise of cities.[16] Most interpretations of social change have been macro-societal approaches, focusing on change at the level of the society as a whole rather than investigating the elements of society.

In contrast, the theory of social evolution that is presented here emphasizes social change through institutional evolution. Institutions are treated explicitly as an analog to the role of genes in Darwinian analysis of biological evolution. Conscious, speaking humans, able to form groups through collective intentionality, are required in order for this process of social evolution to begin. In this institutional-evolution approach to social evolution, the *unit of evolution* is a social group, especially a social institution or a community, for which one may trace the Darwinian changes through the evolution of social institutions based on groups of people.[17] *Variation* appears in the minds of innovators (as individuals and in groups) and in the institutions they propose. *Selection* takes place through "social selection," an analog to natural selection, in which new institutions survive or fail through various processes according to their social fitness. In addition, rather than the dichotomous biological choice of success or failure, institutions may also undergo *regulation* and reform, in which institutional members or social groups modify existing institutions. *Inheritance* depends on preservation and reproduction of information on the institution. For preservation, an *archive* retains information on the institution, where the archive is distributed across human brains and records. For reproduction, *customs* pass information from the archive to the next generation through ritual, rehearsal, and initiation. *Fitness* of any variation in institutions is assessed as strengthening of the relevant social group: such social fitness and strengthening, however, can be assessed according to varying standards. Standards of social fitness are open to debate: fitness can be measured through the number of institutional beneficiaries, their levels of social welfare, or to the welfare of a subgroup, such as an elite. The theory of social evolution focuses particularly on the community as a core institution, though it allows for many other specific institutions, as in production, migration, governance, and exchange, and for emergence of more elaborate institutions with the passage of time. The theory relies on group-based game theory to confirm the effectiveness of group-based decision making.[18] Institutions were small at first; small institutions continue to exist in large numbers, but the modern world also has very large institutions.

---

[16] Theories of social evolution, varying in logic and scale, are discussed in Chapters 9 through 12.

[17] Manning, *History of Humanity*. The theory relies substantially on Raimo Tuomela, *Social Ontology: Collective Intentionality and Group Agents* (Oxford: Oxford University Press, 2013).

[18] Tuomela, *Social Ontology*.

From the standpoint of institutional evolution, the rise of agriculture was the creation of new institutions among communities that had already created other institutions. Emergence of agriculture was more than a single step, in that it involved creating many new social relationships. Agriculture required a collection of institutions dealing with such specific functions as preparing fields, planting, weeding, watering, protecting crops, harvesting, processing harvests, storing, cooking, and preserving seed for the next year. Each of these processes may have its own dynamics (as for the rains, planting, and harvesting). Then, there are questions on who are the beneficiaries of agriculture—producers, consumers, perhaps others? How did the archive for agriculture work—was it the memory of producers, processors, and consumers? How often did the system of agriculture have to be reproduced? (Every year for crops, but multiple years for the land.) How was the "fitness" of agricultural institutions assessed? Were there complaints and needs for revision, as on whether to clear new fields? What evidence do researchers seek from archaeology and social anthropology to clarify the social institutions of agriculture?

Topic 2.3: Social institutions
Question: **What institutions were created to enable agriculture to become successful?**
Theory: institutional evolution.
Evidence: general knowledge on practices of agriculture
Analysis category: Propose a hypothetical relationship
Answer to the Question: (Thought experiment) **Propose a list of institutions sustaining agriculture**. Include the function of each.
Presentation: A list of agricultural institutions, including members, functions, and beneficiaries.
Citation: Manning, *History of Humanity*, 115–18.

Comparing Evolutionary Theories and Levels of Culture

Here are two types of comparison of the three frameworks. Table 2.1 compares the assumptions of biological, cultural, and social evolution in terms of the principles of Darwinian natural selection. The table shows the parallels but also the differences at each level in the three theories, noting terms that are discussed more fully in the text. In addition, there are further differences among the theories beyond those shown in Table 2.1. Biological evolution relies not only on genetic evolution but on epigenetic life-course development, where details on the latter process are still being worked out. Table 2.1 may be helpful in scrutinizing each theory and in comparing them. For instance, the measure of fitness of innovations in each of the three

Table 2.1 Assumptions in various theories of evolution

Evolution type	Biological	Cultural	Social
Unit of evolution	Organism	Organism in dual inheritance	Community or society
Variation: locus	DNA in organism; epigenetics	• DNA in organism • Brain in organism	Brains & institutions in social group
source	Random	• Random • Individual choice	Group choice
Selection: mode	Natural selection	• Personal selection • Inclusive selection • Multilevel selection	Social selection & regulation
Inheritance: retention locus	Inheritance (DNA); epigenetics	• Inheritance (DNA) • brain in dual inheritance	"Archive" (distributed social retention)
reproduction locus	DNA & proteins	• DNA & proteins • Brain in reciprocity & social learning	Customs
Fitness measure	Reproduction of offspring (of gene or organism)	• Personal fitness of individuals • Inclusive fitness of individuals • Multilevel fitness of groups	Social fitness: strengthening of social group (at selected scale)

Note For simplicity, the table assumes neo-Darwinian selection processes, not specifying the more complex processes (for instance, epigenetics) known to be part of the evolution

theories is quite different; it is especially important in considering social conflicts within social evolution. Nevertheless, further challenges arise from this newly-clarified emphasis on the parallels and overlaps of three types of evolution. Do the various processes of evolution reinforce or interfere with one another? Under what circumstances does one or another mechanism become predominant?

A second comparison is on the use and meaning of the term "culture" in the three evolutionary frameworks. Discussion of biological evolution includes few references to culture. On the other hand, the term "culture" is central to the processes of cultural evolution and social evolution, yet it is defined and explored in different ways for each process.[19] That is, the domain of "culture" can now be seen to extend from the most basic level of individual learning to the complex constructions and representations that are created and shared in

---

[19] For a compilation of some sixty definitions of culture, see Alfred L. Kroeber and Clyde Kluckhohn, *Culture: A Critical Review of Concepts and Definitions* (Papers of the Peabody Museum of American Archaeology and Ethnology, Harvard University, Vol. XLVII-No. 1, Cambridge, MA: Published by the Museum, 1952). The definition of culture for studies of cultural evolution was adopted in the 1980s. Boyd and Richerson, *Culture and Evolutionary Process*, 2.

huge societies. It is expected that the understanding of culture at these very different levels will require quite different approaches. In cultural evolution, culture is individual-level learning through trial and error, imitation, or perhaps through instruction. In social evolution, culture is group-level exchange of representations and interpretations, shared and revised through discourse. Group-level culture is created in a world of social institutions; it offers commentary on social institutions and the natural world.[20] In the framework of social evolution, there is a premium on variety within communities, to facilitate innovation and institutions adding to the fitness of each community by expanding its social welfare; hence, migration brings benefits. In the framework of cultural evolution, however, each group builds its fitness and its evolutionary strength through minimizing variety within communities, to enable the group to be more unified in facing competing groups.[21] The distinctions and tensions of culture at various scales are likely to be further revealed and analyzed as the multi-scalar analysis of human evolution proceeds.

Since the word "culture" is applied in many contexts, it is important to specify its use in each framework. I have found it preferable to speak of "individual-level culture," when referring to cultural evolution, and to refer to "group-level culture" when discussing social evolution.

---

[20] Manning, *History of Humanity*, 261–63.
[21] Turchin, *Ultra Society*, 41.

CHAPTER 3

Physical Science and Biological Coevolution

The academic fields of physics, chemistry, and Earth sciences address the basic elements of the physical world. The dynamics of atoms, chemical species, tectonic plates, and the solar system provide the framework for life and continue to influence living things at every step. From that framework, the field of environmental studies extends the physical sciences into the life sciences—it documents the coevolution of living things with each other and their evolutionary interactions with inorganic materials. The lives of animals, plants, and microorganisms have developed complications far beyond the specifics of the physical sciences. Nevertheless, investigation of environmental studies reminds us that those lives begin and end within the constraints revealed by physics, chemistry, and geological and other Earth sciences.

The dimensions of physical sciences that have direct relevance for human history add up to no more than a very small part of the knowledge and research in each of the physical fields. Yet those contributions to history from the corners of the physical sciences make an enormous difference to the understanding of human history. Most basically, they enable us to understand the time frame of human history and its natural environment; they tell us of climate change and its effects. In this concise chapter, I offer a few words on the overall priorities and principles of the fields of physical and environmental sciences, and then turn to specific principles and techniques arising out of each field that are influential in the study of human history.

Physics

The scope of physics in its scale of aggregation is greater than for any other discipline. At the smallest level, what is known as the Standard Model of Elementary Particles includes 17 particles and their fields, in four categories.

These particles combine to form larger particles, though all are too small to be seen directly: electrons and photons are best known. Electric fields arise from the positive and negative charges that control interactions between collections of particles. Magnetic fields are associated with moving charges or currents that circulate between north and south poles. At a larger scale, physics studies the basic particles of protons and neutrons, combined with electrons into atoms. At much the same scale, thermodynamics and statistical mechanics study the interactions of heat and energy within bounded systems. At the scale of visible bodies, physics explores classical or Newtonian mechanics, the gravitational interactions of massive bodies. At the largest scale, the fields of cosmology and astrophysics extend this reasoning to planetary systems, galaxies, and theories of the universe.

Physics organizes its domain in terms of the action of four basic forces and their fields. From the smallest to the largest scale, these are: the weak interactions and strong interactions that govern the relations among the elementary particles; the electromagnetic interactions that center on protons and electrons; and the effects of gravity, attributed to the curvature of spacetime. In the past half-century, physicists have striven to form a "Grand Unified Theory" of these four forces but, to this date, incorporating gravity into their model has eluded them. (It is hypothesized that there may exist a "graviton" as the elementary particle of gravity.) Instruments used in physics include telescopes (measuring light and radio waves) for astrophysics; thermometers and calorimeters for condensed matter physics; measurements of magnetic flux and spectrometers for electromagnetic studies; particle colliders and nanoscale equipment for nuclear and particle physics; and computers for assessing data at all levels.

The disciplines of physics and geology both contributed to determining the age of the Earth. After uranium's radioactivity was discovered, it was understood by 1907 that uranium decayed until it became lead, so the ratio of the two elements in rocks gave an indication of the age of the rock. On this basis, Arthur Holmes proposed 1.6 billion years as age of Earth in a 1913 book. But by 1927, more was known about the many isotopes of uranium and lead, showing three separate "families" of radioactive decay—transforming from uranium-238 to lead-206, from actinium-235 to lead-207, and from thorium-232 to lead-208. (In all of these cases, there were also ratios of helium to the heavy elements.) Estimates gradually became more consistent: geological analysis by 1937 showed the oldest known rocks were 2 billion years old. By 1945, physicists used mass spectrometers to measure the relative abundance of uranium isotopes to give a definitive age of 4.5 billion years for Earth's age. Thereafter, geologists found early meteorites that confirmed the earliest known date as 4.5 billion years ago.

> Topic 3.1: Using physics to date the Earth
> *Question:* **How do we know the age of the Earth?**
> *Theory:* physical behavior of isotopes, chemical composition, geological composition.
> *Evidence:* mineral samples; weights and half-lives in families of elements and decay (uranium, actinium, thorium)
> *Analysis category:* Estimate parameters
> *Answer to the Question:* **Physics and chemistry of uranium and the geology of early stones led to estimates of Earth's age.** Estimates of 2 billion years in 1937 and 4.5 billion years in 1945.
> *Presentation:* Estimates show earliest known date, including ages of all the intervening geological strata of Earth.
> *Citation:* Holmes, *The Age of the Earth* (1937), 99–100, 138–44.

CHEMISTRY

The academic field of chemistry addresses only a small portion of the physical world explored by physicists yet analyzes that scale in great detail. Chemistry addresses atoms of the various chemical elements, but it also addresses the principal subatomic particles (protons, neutrons, electrons, and photons), as well as molecules and compounds. Each element has a nucleus with an atomic number (the number of positively charged protons); an atomic weight (the number of protons plus the number of uncharged neutrons); and relatively small, negatively charged electrons in orbits around the nucleus. Many elements have multiple isotopes because they have different numbers of neutrons. While most isotopes are stable, some are unstable, decomposing and emitting radiation: radioactive isotopes are known by their "half-life," the amount of time it takes for half of the original atoms to decay.

Atoms combine into molecules, electronically neutral groups of two or more atoms. Chemical compounds are molecules composed of two or more chemical elements. Chemistry gives great attention to the chemical bonds that link atoms into molecules. At varying levels of temperature and pressure, both elements and compounds may take forms of solids, liquids, or gases. Chemists give substantial attention to chemical reactions, in which chemical compounds and elements interact with one another, transforming one set of substances to another set. Fire is a chemical reaction in which oxygen reacts with other compounds or elements, especially carbon, resulting in the generation of both heat and gases.

The main subfields in chemistry (working from the smallest to the largest scale) are physical chemistry, inorganic chemistry, organic chemistry, and biochemistry. Physical chemists work closely with physicists in studies that center on quantum mechanics in the analysis of atoms, also including the thermodynamics (heat and energy) of atoms and molecules. Inorganic chemistry explores the chemical compounds of metals and salts that do not rely on bonds between carbon and hydrogen. Organic chemistry is the study of carbon-hydrogen-based compounds, often large compounds that are the basis of life. Biochemistry analyzes very large organic compounds, especially in their functions in living organisms. Because of the importance of carbon-based or organic molecules in living systems, organic chemistry and biochemistry have persisted as major fields within chemistry.

Both physics and chemistry are laboratory sciences more than historical sciences, in that their processes are not greatly influenced by the passage of historical time. Theory in chemistry overlaps greatly with that of physics: it includes atomic theory, quantum mechanics, relativity, thermodynamics, statistical mechanics, and macrosystems. Chemists are especially concerned with the specific characteristics of the 118 known elements (94 of which occur naturally on Earth) and with their interactions in molecules and as ions (atoms or molecules with positive or negative electric charges). Instruments relied upon in chemistry include equipment for chromatography, spectrometers, the electron microscope, DNA sequencers, and nanoscale instruments.

In 1939, researchers found that radioactive carbon-14 (heavier than the naturally occurring carbon-12) was produced when cosmic rays struck atmospheric nitrogen. From the atmosphere, carbon-14 descends into land and water, and is incorporated into carbon-based plant and animal matter. Carbon-14 has a half-life of 5730 years, so that it could still be detected after nine half-lives, a period of 50,000 years. Comparison of the ratio of carbon-14 to carbon-12 in the sample to a standardized table provided a basis for calculating the age of the sample. Initial studies assumed a constant rate of formation of carbon-14 in the atmosphere. But studies of tree rings, which form a coherent record back to about 10,000 years ago, showed how the carbon-14 concentration changed in time, affecting derived ages. Conversion tables now estimate the relationship between a sample's carbon-14 concentration and calendar years. The ratio of carbon-14 to carbon-12 in a sample, initially measured by beta-rays (electrons) emitted, is now traced by using a mass spectrometer to determine the sample's ratio of the two isotopes. Willard Libby won a 1960 Nobel Prize for developing the analytical technique.

Anthropologist Tom Dillehay, excavating the site of early human settlers at Monte Verde, Chile, located samples of bone and burned wood at a deep level. He extended his analysis to numerous small and separated but nearby carboniferous samples, in effect proposing a new rule for archaeological analysis. This method enabled him to argue that humans had settled at Monte Verde by 18,500 years ago.[1]

---

[1] Tom D. Dillehay et al., "New Archaeological Evidence for an Early Human Presence at Monte Verde, Chile," *PLoS One* (November 18, 2015): 1–5. DOI: 10.137.

> **Topic 3.2: Radiocarbon dating of human remains**
> *Question:* **How does radiocarbon analysis determine the date of human settlement in South America?**
> *Theory:* chemistry; archaeology
> *Evidence:* archaeological data from Monte Verde site, Chile
> *Analysis category:* estimate parameters
> *Answer to the Question:* **Radiocarbon analysis of samples that were scattered (but close to each other) show earliest human occupation of Monte Verde site at 18,500 years ago.**
> *Presentation:* Dillehay presents arguments for treating small and separated pieces as parts of a single site, extending radiocarbon practice, to confirm an early human settlement date.
> *Citation:* Tom D. Dillehay et al., "New Archaeological Evidence for an Early Human Presence at Monte Verde, Chile," *PLoS One* (November 18, 2015): 1–5. DOI: 10.137.

EARTH SCIENCES

The academic field of Earth sciences grew out of its origins in geology, stretching across a wide range of related topics. Its initial advances came through studies of stratigraphy, the layering of geologic forms deposited over time. The field now works at scales from the molecular level to the analysis of the Earth and other planetary bodies. Earth sciences is a historical field, as its processes must be identified in terms of time. One may define the scope of Earth sciences in terms of its various spheres: the atmosphere that surrounds the planet; the geosphere that includes the rock, soil, and interior of the Earth (including the geosphere's upper levels, the lithosphere); the hydrosphere, especially the oceans but also the lakes, rivers, and streams; and the biosphere, all the living inhabitants, large and small. Earth sciences began with mineralogy and topography, and then extended to the study of sediments, oceanography, tectonics, climatology, and astronomy.

Within Earth sciences, studies of climate currently hold the most attention. (Meteorology, the short-term study of weather, has remained distinct from long-term climate studies.) An authoritative textbook divides climate studies into those of tectonic-scale climate change (the last several hundred million years, in which one traces long-term climate in terms of the quantity of carbon dioxide and the movement of tectonic plates); orbital-scale climate change (the last 400,000 years, with climate changes in response to variations in solar radiation, monsoon systems, and periodic glaciation); deglacial climate change (since the Last Glacial Maximum some 20,000 years ago); and human influences on climate within the past 10,000 years.[2]

---

[2] William F. Ruddiman, *Earth's Climate: Past and Future*, 3rd ed. (New York: W. H. Freeman and Co., 2014).

Methods for dating samples formed in the past, used more commonly in the Earth sciences than in other fields, depend on several types of technology, of which radioactive technology is best known. Incremental dating includes dendrochronology of tree rings, the strata within carefully retrieved ice cores, analysis of the varves that are the annual additions to sedimentary rocks, and the strata that can be identified in the remains of lichens. In luminescence dating, sediments of quartz, diamond, and feldspar, or buried and "fired" objects such as pottery and cooking stones, contain excited electrons that had been released by ionizing radiation from molecules of potassium, uranium, etc., but caught in crystal lattices. Application of heat or light in a lab releases and measures the electrons that accumulated over time. This technique can give dates for burial of the materials ranging from 100 to 350,000 years ago. Additional techniques for dating rely on paleomagnetism, magnetostratigraphy, and chemostratigraphy.

Ice cores give information on air temperature and air chemistry over time. Dates are established by annual layers in ice cores—measured by geochemistry, ash layers, and electrical properties of layers. Air temperature within each layer is calculated from a ratio of oxygen isotopes, comparing the rare ^{18}O to common ^{16}O, known to correlate with past temperature. Atmospheric greenhouse gases, methane and carbon dioxide, can also be measured in ice-core layers. Ice cores of up to 3000 meters in depth in Antarctica provide information on the past 800,000 years.

> **Topic 3.3: Earth sciences**
> *Question:* **How do ice cores document weather 10,000 years ago?**
> *Theory:* Earth science; chemistry
> *Evidence:* ice cores
> *Analysis category:* estimate parameters
> *Answer to the Question:* **Sample dates are given by ice-core stratigraphy; temperature is calculated from $^{18}O - ^{16}O$ ratios; proportions of methane and CO_2 are measurable.**
> *Presentation:* Typical graph from Vostok station in Antarctica shows measures of 5 variables over 400,000 years.
> *Citation:* "Ice Core Basics," http://www.antarcticglaciers.org/glaciers-and-climate/ice-cores/ice-core-basics/.

BIOLOGICAL COEVOLUTION IN THE ENVIRONMENT

The expanding discipline of environmental science combines a mix of physical, biological, and information sciences: atmospheric sciences, ecology, environmental chemistry, geosciences, and systems analysis. Environmental science has a substantial overlap with Earth sciences. Research focuses on the movement of materials and energy through living communities; successional development of ecosystems; distribution of organisms and biodiversity. Subfields focus on soil, freshwater biology, marine biology, parasitology, and population dynamics. Research takes place in laboratories but also in fieldwork worldwide. Field instruments aid in sampling air and water; lab instruments include spectrometers for analysis of field samples.

The issue of coevolution has become increasingly central to environmental science. Coevolution refers to the evolution of at least two species, which occurs in a mutually dependent manner. It functions by reciprocal selective pressures on two or more species. Formal study of coevolution began with the study of insects and flowering plants but it has since expanded to processes including sexual reproduction and infectious disease. Well-known examples include predator-prey, host-parasite, and other competitive relationships between species. But coevolution can also provide mutual benefit to each of the interacting species. In each case, coevolution functions by reciprocal selective pressures on two or more species.

In generalizing results of the first few decades of study, John Thompson proposed four central points about coevolution, as follows: "(1) Complex organisms require coevolved interactions to survive and reproduce. (2) Coevolution builds species-rich ecosystems. (3) Coevolution takes multiple forms and generates diverse ecological outcomes. (4) Interactions coevolve as constantly changing geographical mosaics." In earlier work, Thompson gave a thorough description of the raw materials for study of coevolution.[3]

In this exercise, you are asked to review five concise examples of the interaction of a pollinating species with the flowing plants to which each is linked. All five examples are drawn from the flora and fauna of Washington State in the northwest United States. To develop your understanding of the working of coevolution, you are asked to write descriptions of two of the coevolutionary processes that result in successful pollination of flowers.

Topic 3.4: Biological coevolution
Question: **What are two examples of successful relationships between a type of pollinator and a type of flower?**
Theory: ecology
Evidence: patterns of coevolution among multiple species.
Analysis category: relationship
Answer to the Question: **In the selection by Arnett, five cases of pollinators (such as bees or flies) are linked to specific flower types. Select the two of most interest to you and describe details of the relationship between pollinator and flower, including why you believe it is successful.**
Presentation: Select two out of the five cases presented and describe the relationship between the pollinator and the flower.
Citation: Joe Arnett, "Coevolution and Pollination."

---

[3] Joe Arnett, "Coevolution and Pollination," *Washington Native Plant Society* (2014), http://www.wnps.org; John N. Thompson, "Four Central Points About Coevolution," *Evolution Education Outreach* 3 (2010): 7–13; and John N. Thompson, "The Raw Material for Coevolution," *Oikos* 84 (1999): 5–16.

CHAPTER 4

Systems and Information Science

Substantial developments unfolded in physical, biological, and social sciences in the early twentieth century—including logical, mathematical, and empirical formulations of theories. Social science examples included population studies, input-output economic analysis, national income analysis, and other fields. Perhaps in response, mathematicians and mathematically oriented scientists focused on developing abstract yet general and coherent models of systemic behavior. The term and concept of "system" had long been used—the term "solar system" was coined in the eighteenth century—yet "system" had been a vague metaphor for complexity rather than a specific formulation. New and more specific work of the 1920s—by Sewall Wright, Alfred Lotka, and Ludwig von Bertalanffy—pointed toward the formalization of systems logic. More broadly, the expansion of quantitative analysis in numerous disciplines pointed, eventually, to the need for a broad, interdisciplinary focus on logical and quantitative data structures.

With the end of World War II, several analytic disciplines arose, in which empirical details were subordinated to logical structure. Multiple advances in logical and quantitative modeling appeared; they were soon followed by the development of advanced computing capacity.[1] The numerous and varied initiatives gradually settled down into the broad categories of systems analysis and information science. This chapter emphasizes some dimensions of that broad development that were to have direct implications for the study of human history in both short-term and long-term perspectives. The focus here is initially on systems because of their conceptual importance to historical methods, as with multiple scales and the interplay of subsystems. In addition, the full range of information science is of importance for studies in human

---

[1] For a concise review of the expansion of computing capacity, see Paul E. Ceruzzi, *Computing: A Concise History* (Cambridge, MA: MIT Press, 2012).

history: the latter portion of this chapter identifies some conceptual and practical advances that rely on computers and data systems.

General Systems

A system is a complex of interacting elements. The most persistent formal theorist of systems was Ludwig von Bertalanffy, a Swiss researcher who conducted his initial work in the early days of biochemistry.[2] He defined systems as closed or open: closed systems are self-contained; open systems ingest and extrude matter and information. Systems are generally self-regulating: closed systems self-correct through feedback while open systems tend toward a steady state. System thinking is both part-to-whole and whole-to-part thinking about making connections between the various elements so that they fit together as a whole. Communities, confederations, societies, networks and combinations of these are social subsystems, each containing functional sub-subsystems that perform their designated tasks.

Open systems and tendencies toward a steady state. Open systems—in which internal subsystems interact—tend to persist, grow, develop, and reproduce. The logic of a closed system—such as a thermostat maintaining temperature in a house through feedback—has been explored in depth. The logic of open systems—such as a neighborhood or a school—still requires further study. In an open system, energy is transferred not only within the system but between the system and its surroundings. A closed system has one type of transfer of matter and energy: within the system. An open system has three types of transfer of matter and energy: input to the system, transfer within the system, and output from the system. Within the system, subsystems handle the origin, regulation, and exchange of components. A remarkable characteristic of open systems is that their purposeful work can develop to higher levels of order and organization, thus apparently violating the Second Law of Thermodynamics. An open system is never in equilibrium, but it tends to adjust toward a "steady state," in which "the system remains constant in its composition, in spite of continuous irreversible processes, import and export, building-up and breaking-down, taking place."[3]

In open chemical and biological systems, chemical reactions and biological metabolism, within the system, perform tasks that enable the system to survive and perform its function. The interaction of these functions is a key issue in open-systems analysis. The system and its subsystems must be renewed, repaired, and regulated. Cells within organs of living things may have rapid turnover time, but the organ persists. Physiological and hydrodynamic models show the operation of active transport among subsystems.

---

[2] Von Bertalanffy first published on systems in 1928; his publications expanded from 1940 and his book-length overview appeared in 1969. Ludwig von Bertalanffy, *General System Theory* (New York: George Braziller, 1969).

[3] Von Bertalanffy, *General System Theory*, 142.

Maintaining a steady state requires complex interactions of internal subsystems with each other and with both import and export of matter and energy. In an interesting relationship, Bertalanffy notes that, as animals get larger, their metabolic weight (and therefore the input of nourishment) increases not in a simple ratio to their weight but to the growing surface area of the animal.[4]

> **Topic 4.1: General systems**
> *Question:* **How do ingest and output stay in balance in open systems?**
> *Theory:* General systems theory
> *Evidence:* workings of a system including chemical reactions and biological metabolism
> *Analysis category:* Propose a hypothetical relationship
> *Answer to the Question:* **Reactions and metabolism interact within the system, also reaching beyond the system to gather inputs and deposit outputs.** An open system cannot equalize input and output but can adjust output and input toward each other.
> *Presentation:* Discussion of system functioning, input and output, showing the energy expended in maintaining a steady state.
> *Citation:* von Bertalanffy, *General System Theory*, 139–49.

Gaia Within the Earth System

It was not until the combination of multiple areas of geologic research that the notion of the Earth system arose to provide a more comprehensive view. The term "Earth system" gradually arose within the Earth sciences to refer to Earth's interacting physical, chemical, and biological processes. The system consists of the land, oceans, atmosphere, and poles. It includes the planet's natural cycles—the carbon, water, nitrogen, phosphorus, and other cycles—and deep Earth processes. Life too is an integral part of the Earth system.[5]

Gaia theory, developed by inorganic chemist James Lovelock and microbiologist Lynn Margulis, proposes that all organisms and their inorganic surroundings on Earth are closely integrated to form a single and self-regulating complex system, maintaining the conditions for life on the planet. The initial insight, from Lovelock's comparison of cold Mars, hot Venus, and just-right Earth, was that an additional process maintained Earth's systems within conditions that could support life. As new evidence appeared, it became clear that, for the past 500 million years, Earth's average temperature had

---

[4] Von Bertalanffy identifies this as an *allometric* relationship. Von Bertalanffy, *General System Theory*, 171ff.

[5] Ruddiman, *Earth's Climate*; James Lovelock, *Gaia, a New Look at Life on Earth* (Oxford: Oxford University Press, 1979).

remained between −1 °C and +8 °C; the ocean's salinity had remained at roughly 3.4%; and atmospheric oxygen, currently 21% of the atmosphere, had remained between 15 and 25%. Lovelock and Margulis located overlapping mechanisms for all three of these equilibria. For atmospheric temperature, greenhouse gases (carbon dioxide and methane) came from volcanoes, from weathering of carbonate rocks, and from forest fires. In contrast, photosynthesis, especially in forests, absorbs carbon dioxide and water, producing oxygen and sugar.[6]

Lovelock found that methane and oxygen are in surprisingly high quantities in the atmosphere, as they would be expected to react to form carbon dioxide and water vapor. They can only be at their known levels if methane is introduced to the atmosphere and if there is a way to replace the oxygen used up in oxidizing methane. Similarly, the amounts of nitrous oxide, ammonia, and molecular nitrogen in the atmosphere are surprising—something is adding nitrogen to the atmosphere, since it is most stable as nitrate ions dissolved in the oceans.

Topic 4.2: Gaia
Question: **What conflicting forces brought temperature equilibrium in Gaia?**
Theory: Gaia theory within Earth sciences
Evidence: chemical proportions of gases in the atmosphere, compared with what one would expect if living things did not offset the balance.
Analysis category: Propose a hypothetical relationship
Answer to the Question: **Methane, oxygen, and nitrogen must be introduced to the atmosphere by processes of living organisms.** Otherwise, quantities of these gases would fall to near zero and Earth's temperature would fluctuate wildly.
Presentation: Lovelock tells the story of his discovery of the logic of Gaia, with a few key paragraphs on his deduction about the way organisms function in Gaia.
Citation: Lovelock, *Gaia: A New Look at Life on Earth*, loc. 318–431.

LIVING SYSTEMS

Biological systems function at multiple scales, from the cellular to that of the organism and on to the scale of herds. James G. Miller, in his highly structured 1978 typology and analysis, showed the analogies among all these scales of living systems. He emphasized that living systems must perform many functions and, thus, are composed of relatively complex subsystems. In

---

[6] James Lovelock, *Gaia, a New Look at Life on Earth* (Oxford: Oxford University Press, 1979); for an update and reconsideration, see Lovelock, *The Revenge of Gaia: Earth's Climate Crisis and the Fate of Humanity* (New York: Basic Books, 2007).

arguing for the unity of the functions of life, Miller argued that each of seven scales of living systems relies on a parallel set of twenty subsystems that process inputs, throughputs, and outputs of various forms of matter-energy and information, each performing a function necessary for the survival and reproduction of the system.[7] Two of these subsystems—reproducer and boundary—process both matter-energy and information. Eight of them process only matter-energy. The other ten process information only. In the long history of evolution, a new systemic scale occasionally emerges. Such was the case, long ago, as organisms arose from cells and as organisms gave rise to groups.[8]

Miller's seven scales of living systems, each encompassing the preceding scale, are modified slightly here. In listing them in order, I indicate in brackets the different labels and modified definitions that I give to two of them: cell, organ, organism, group [family], organization [institution], society, and supranational system.[9] While all of these are living systems, the first four are clearly biological systems, while the last three are arguably social systems. Since all are clearly living systems, Miller did not worry much about this difference, nor did he pay much attention to the difference between biological reproduction and social reproduction. Miller's focus was on structure and function of living systems, more than on reproduction, evolution, or change. But his analysis does provide a comprehensive overview of the scales of existence, showing parallels, distinctions, and cross-scale linkages in units from the cellular to the macro-societal scales. Each of Miller's scales was conceived as a system with its subsystems, in systemic interplay with the scales above and below. The scales developed in a certain order. While cells were first to evolve, a second scale, organs, could not form without organisms to link them, and organisms could not function without organs—they thus evolved at much the same time.[10] Groups or families, consisting of a number of organisms, long existed as a separate scale. In changing the labels of the fourth and fifth scales,

---

[7] James G. Miller, *Living Systems* (New York: McGraw-Hill, 1978), xxiv–xxvi, 30–33. See also G. A. Swanson, "James Grier Miller's Living Systems Analysis (LSA)," *Systems Research and Behavioral Science* 23 (2006): 263–71.

[8] John Maynard Smith and Eörs Szathmáry, *The Major Transitions in Evolution* (Oxford: Oxford University Press, 1995).

[9] For humans, I treat the "group" as referring to the biological family. For other species, the biological "group" can vary in size and structure, as with flocks of birds or schools of fish. As defined by Miller, "A group is a set of single organisms ... which, over a period of time or multiple interrupted periods, relate to one another face-to-face, processing matter-energy and information." Miller, *Living Systems*, 515.

[10] Miller, *Living Systems*, 315–16, 357.

I emphasize that families exist primarily at a biological and unconscious scale, while institutions exist primarily at a social and conscious scale. Necessarily, the two scales interact with each other.[11]

For the case of an individual human organism and its subsystems, the twenty categories listed here provide the framework for analysis.[12] This list of subsystems provides a useful subdivision of biological activities into the organism's principal tasks: reproduction of the system, processing matter and energy, and processing information:

> For the organism's system as a whole. *(1) Reproducer. (2) Boundary.*
> For treatment of matter and energy. *(3) Ingest. (4) Distributor. (5) Convertor. (6) Producer. (7) Storage. (8) Extruder. (9) Motor. (10) Supporter.*
> For treatment of information.[13] *(11) Input transducer. (12) Internal transducer. (13) Channel and net. (14) Timer. (15) Decoder. (16) Associator. (17) Memory. (18) Decider. (19) Encoder. (20) Output transducer.*

The links among the subsystems are not detailed here, but the reader should be able to fill in many gaps by thinking about the well-known processes of human life. Within each of these principal tasks, one or more subsystems focuses on relations beyond the system, through ingest, extrusion, and maintaining the system's borders. It may be surprising to see how many of the subsystems are devoted to processing information—both for the scale of the organism as shown here and down to the scale of the cell—but these subsystems provide a reminder of the degree to which the existence of life itself depends on information.

Biological subsystems are neither neatly packaged nor discrete. For instance, the *output transducer*, to express the individual's choices and decisions to the environment, uses numerous agents—including hands, feet, and emotions, plus voice (which became more important once humans developed language).[14] In addition, a single agent can contribute to several subsystems—in the human organism, the mouth contributes to the subsystems of *ingest, extruder,* and *output transducer.* Subsystems, therefore, consist of all the elements that combine to perform each specific function: it is important not to oversimplify them.

---

[11] In addition, I treat communities as institutions. Such institutions are building blocks from which all larger scales of human society are constructed. For details, see Manning, *A History of Humanity*.

[12] Miller, *Living Systems*, 30–33, 93–103. On subsystems for human organisms, see 363–65.

[13] The "timer," not included in the original 1978 version, was added later. G. A. Swanson, "James Grier Miller's Living Systems Analysis (LSA)," *Systems Research and Behavioral Science* 23 (2006): 267.

[14] Further, as Miller pointed out, the functions of certain subsystems are dispersed downward to the next scale in the biological hierarchy, yet still serve as an overall subsystem "A living system does not exist independently if its decider is dispersed upwardly, downwardly, or outwardly." Miller, *Living Systems*, 32.

> **Topic 4.3: Living systems**
> *Question:* **How do subsystems keep a human organism alive?**
> *Theory:* Living systems; biology
> *Evidence:* subsystems of a human individual
> *Analysis category:* Propose a hypothetical relationship
> *Answer to the Question:* **Reproduction systems create the next generation. Matter-energy systems process, metabolize, and expel remnants of foodstuffs and oxygen. Information systems locate food and other resources, and direct the interconnections of all internal processes.**
> *Presentation:* Describe examples of conscious but especially unconscious steps in information, matter-energy, and reproduction—including ingest, internal operation, and extrusion.
> *Citation:* Miller, *Living Systems*, 361–68.

HUMAN SYSTEM

The *Human System* is a historically formed totality of the processes carried out by humans. It took form through initial links among nearby communities; thereafter, the system expanded progressively to encompass links among confederations and societies. As hierarchies emerged, the Human System incorporated links among hierarchical scales. The Human System, at the scale of all humanity, is at a scale larger than those identified by Miller.[15] Still, the Human System was more than the encompassing envelope: it included the vertical and horizontal links interacting at every scale down to the local. It is a living system in the terms of Miller.

Human communities, starting with local communities, are *open* systems: they interact with their environment and they also depend closely on neighboring communities. This open-community structure has been essential for the multiple scales of communities: local, language-based communities; towns as they developed; ethnicities and monarchies; regional groupings; and states (including monarchies, empires, and nations). Because community systems are open, the subsystems that perform functions for sustaining them are often *distributed* among communities or institutions. For instance, collection of food can be distributed across communities of foragers, herders, fishers, and farmers. Subsystems, therefore, consist of all the elements that combine to perform each specific function.[16]

---

[15] I argue that the Human System extends Miller's criteria for living systems to a new level. Miller's largest category was that of the "supranational," including empires. For Miller, "in an empire, the decider of one society constituted an echelon above those of a number of other societies." I prefer to eliminate "supranational" as a separate stage, including empires within the stage of societies, but I also prefer to treat the Human System as a separate and species-wide stage. Miller, *Living Systems*, 907–908.

[16] Subsystems are too often misunderstood as coherent and neatly bounded. This common error in categorization is repeated when observers exaggerate uniformity within groups and also differences between groups.

When human social evolution began, multi-family communities needed to form in order to sustain speech; families ended up being restructured in order to function with speech and to find their place within communities. At a later stage, societies formed at a larger scale as humans took up productive activities and developed complex governance. But as with the inevitable inerplay in the formation of organs and organisms, societies could not form without institutions (in which people shared specialized tasks or objectives). The initial institutions and communities existed for many thousand years before they were combined to form the basis for constructing societies. Once societies had been constructed, they underwent transformation and perhaps expansion in interaction with the continuing evolution of social institutions.

The Human System has grown to immense size and complexity. It is not guaranteed that all of its subsystems will function or fit together properly. Here you are asked to draw on your experience of the twentieth and twenty-first centuries (in social, economic, health, cultural, or other affairs), to express that experience in terms of subsystems and their links, and then propose some examples of systemic malfunction, potential negative effects, and possible corrective steps.

Topic 4.4: Human system
Question: **What problems in subsystems might cause damage to the Human System?**
Theory: institutional evolution; living systems
Evidence: examples drawn from functioning of twentieth- and twenty-first-century Human System
Analysis category: Propose a hypothetical relationship
Answer to the Question: (Thought experiment) **Identify and describe subsystems that might misfunction, showing potential negative effects and possible regulation.** Use the same list of subsystems as for Topic 4.3.
Presentation: Summarize examples of subsystems, possible malfunctions, and regulation that might improve functioning.
Citation: Manning, *A History of Humanity*, 225–34.

World-Systems

The concept of World-System, as developed by Immanuel Wallerstein, was to explain the rise of capitalism of the seventeenth to nineteenth centuries through systemic interaction of a European center, colonial peripheries, and intermediate semiperipheries.[17] As elaborated by others, this systemic framework and its dynamics (including cyclical economic and political changes) have been applied in other times and at other scales.

---

[17] Immanuel Wallerstein, *Modern World-System*, Vol. 1 (New York: Academic Press, 1974).

Christopher Chase-Dunn and Thomas D. Hall developed the most comprehensive extension of World-Systems theory to apply to numerous historical situations over the past 5000 years.[18] The analysis distinguishes four overlapping types of networks—information, exchange, political-military, and bulk goods. Within this framework, the analysis traces the expansion of political-military and exchange networks from Mesopotamia through Afro-Eurasian links to the modern World-System. Analysis focuses on growth and interplay among its units, with attention to semiperipheral initiative, but without a major focus on subsystems.

A key early step in the expansion of this network was the rise of the Akkadian Empire in Mesopotamia. Sargon took power in Akkad in about 2340 BCE and conquered all of the Sumerian cities. Factors giving advantage to Akkad, as cited in the literature, include "spread" and "backwash" effects, recruiting nomadic warriors, improved archery, class conflict, and weakness in core states.

Topic 4.5: World-System
Question: **What does World-Systems theory explain about Akkadian conquest of Sumeria?**
Theory: World-Systems theory
Evidence: narrative of the strategy and wars of conquest in which Mesopotamian city-states were formed into an empire
Analysis category: verify a hypothesis
Answer to the Question: **Akkad, a semiperipheral state, combined military, balance of power, and economic strategies to conquer.**
Presentation: Discussion of numerous interpretations of Akkadian conquest, identifying their common elements within the framework of semiperipheral initiative.
Citation: Chase-Dunn and Hall, *Rise and Demise*, 78–79, 84–89.

INFORMATION SCIENCE

Information science congealed as a disciplinary focus in the last half of the twentieth century. It relied heavily on the expanding capacity of electronic computers but set computation in the context of analyses of information generally. Its fields include computer science, artificial intelligence, archival and library science, statistics, ontology, databases, simulation, and visualization. The field is still expanding and organizing itself. Applications and linkages to other disciplines are growing rapidly, including to health sciences, linguistics, management, museology, digital humanities, public policy, and social sciences. The discussion here touches on some but not all of its dimensions.

---

[18] Christopher Chase-Dunn and Thomas D. Hall, *Rise and Demise: Comparing World-Systems* (Boulder, CO: Westview Press, 1997).

Archives, Cataloguing systems. Archives, databases, and cataloguing systems have developed impressively in many fields and have become remarkably accessible since the rise of the Internet. The Human Areas Research Files (HRAF) brought an early advance in anthropology beginning 1947; the archive has been updated periodically.[19] The largest archive of medical and health research is the US National Library of Medicine, which has grown under various titles since 1836.[20] In 2000, this institution opened PubMed, an open-access, online catalog of titles and abstracts for all medical and health literature, with more than 30 million citations and abstracts of peer-reviewed biomedical literature.

You are invited to explore PubMed through its User Guide and to prepare a brief list of search topics of interest to you. An example of a search term you can enter is "Tianyuan" or "Tianyuan Cave." This will lead you to articles on the remains of a man who lived 40,000 years ago and was found in a cave near Beijing. PubMed, once you have built up some practice, can be extremely useful. On the other hand, topical searches on Google are easy to submit and often bring up very useful results.

Topic 4.6: Archives
Question: **How can one locate topical resources within PubMed?**
Theory: Library science, health sciences
Evidence: academic publications in health science and related fields
Analysis category: define an ontology
Answer to the Question: (Thought experiment) **Based on the online PubMed User Guide, develop a classification of topics of interest to you.**
Presentation: Identify relevant sections of the User Guide, note key terms and definitions, and prepare a one-page guide to finding publications of interest to you.
Citation: PubMed home, pubmed.ncbi.nlm.nih.gov/pubmed; PubMed User Guide, pubmed.ncbi.nlm.nih.gov/help.

Ontology. Ontology is the categorization of groups and subgroups of items within a domain. Domain ontology is the categorization of items specific to an area of study. An upper ontology is a set of maximally general categories, such as continents within a geographical ontology.

---

[19] Human Relations Area Files, https://hraf.yale.edu.

[20] National Library of Medicine (www.nlm.nih.gov): the Surgeon General's library, formed 1836 in Washington DC, grew steadily from 1865. It became the National Library of Medicine in 1962. Its electronic version began with Medline in the 1970s, which became the open-access PubMed in 2000. Another major resource is the Wellcome Library in London, at wellcomelibrary.org.

Professional groups with a strong interest in sharing a common ontology for their area of work have developed domain ontologies including BORO Business Objects Reference Ontology; CIDOC Conceptual Reference Model, for cultural heritage; ISO 15926, International Standard for representation of plant life-cycle information; and UFO, Unified Foundational Ontology. As another domain ontology, a world-historical gazetteer, released in 2020, lists global historical places and links them to information on the places: see whgazetteer.org.

In this exercise, you are requested to categorize types of places in your home town. You may list individual places, but the main point is to identify the types of places and to suggest something about the relationship of the different places. You will encounter choices in categorizing places; then, you will make the choices; finally, you will look at the results of the choices you have made and see what you have learned. As an example of a very simple gazetteer, see the list of towns in Galicia (now Poland and Ukraine) remembered by Jewish emigrants, at www.geshergalicia.org.

Topic 4.7: Ontology
Question: **What should be the structure of a geographical ontology for your home town?**
Theory: Ontology
Evidence: Your own information on places in your home town.
Analysis category: define an ontology
Answer to the Question: (Thought experiment) **Identify places and types of places in your home town, and prepare an ontology.** It should organize places by type and scale, showing how the types are linked to one another.
Presentation: A one-page list of categories (with an example for each category) showing their links.
Citation: World-historical gazetteer, www.whgazetteer.org.

Databases. Databases or datasets, generally stored in electronic form, are the results of collection, organization, displaying, analysis, interpretation, and presentation of data. They are usually organized as two-dimensional spreadsheets, with variables listed in one axis, cases observed on the other axis, and with the observations of data in the cells linking variables and cases. Data may be collected from historical research in past documents, from current data such as online records, or from data mining and sampling of data from large data archives. Data must be verified for accuracy and consistency before analysis. If there are missing observations on variables, one faces choices on whether to neglect the variable or to simulate the missing data. Datasets may be constructed to fit with variables in a theory in order to test theoretical relationships. Analysis of datasets can explore the nature and significance of relationships among variables.[21]

The Seshat database has been constructed for comparative and global analysis of societies since the beginnings of agriculture, expanding earlier such datasets.[22] In this analysis of numerous societies over time, the researchers ask whether the societies largely resemble each other, or whether they tend to fall into two or more categories. The method of analysis is *principal component analysis*, a statistical technique comparing multiple observations and using an algorithm to simplify them into one or more categories. In this case, most of the variation in the database can be explained by a single component, i.e., as variations on a single model. As the authors argue, "We found that different characteristics of social complexity are highly predictable across different world regions."

Topic 4.8: Databases
Question: **Do human societies from around the world exhibit similarities in the way that they are structured? Do they show commonalities in the ways that they have evolved?**
Theory: authors sought to select data that were as "theory neutral" as possible
Evidence: for polities selected from 30 "natural geographic areas" of the world, data were collected on 51 variables, aggregated into nine "complexity characteristics."
Analysis category: verify a hypothesis
Answer to the Question: **Yes. "Our analyses revealed that these different characteristics show strong relationships with each other and that a single principal component captures around three quarters of the observed variation."**
Presentation: The article describes the process of sampling (among regions and variables), an ontology for aspects of human society and statistical procedures (focusing on principal component analysis).
Citation: Seshat, http://seshatdatabank.info; Turchin, et al., "Quantitative Historical Analysis Uncovers a Single Dimension of Complexity That Structures Global Variation in Human Social Organization," *PNAS* 115 (2): E144–E151.

---

[21] For instance, the Caribbean inequality database, discussed earlier in Topic 1.5, has a different structure and a different purpose than Seshat. For further considerations on cross-disciplinary datasets, see Patrick Manning and Sanjana Ravi, "Cross-Disciplinary Theory in Construction of a World-Historical Archive," *Journal of World-Historical Information* 1, 1 (2013): 16–39.

[22] Seshat: Global History Databank (seshatdatabank.info).

CHAPTER 5

Behavior of Individuals, Groups, and Networks

This chapter reviews the explorations and identifications of individual and group behavior among humans, especially as related to human evolution; by extension, it discusses individuals and groups in networks. In biological studies, Darwin focused his analysis on natural selection for individual organisms—the variations in their characteristics that made them more or less fit for survival, as indicated by the number of their offspring. Similarly, individual behavior has provided the framework for most of the detailed study of human behavior in economics, sociology, and psychology. Nevertheless, both animals and humans are known to exhibit behavior not only as individuals but also in groups, as indicated in the herds of bison, packs of wolves, schools of fish, flocks of birds, and especially the hives of bees and the nests of ants.[1] Further, the practice of speech among humans and, with it, the sharing of conscious reasoning, has brought prominence to group behavior and interest in analyzing it. Not yet fully resolved is the recurring question of whether the descriptive evidence of groups requires that groups be studied at the analytical level. Is group behavior a logical extension of individual behavior? Or is group behavior a different type of behavior that can be explained only with the addition of additional factors, relying on a different theory?

For several centuries, writers have told tales of both individual and group behavior, though the logic of individual behavior has been advanced with greater specificity than that of group behavior. Thomas Hobbes and John Locke, followed by Jean-Jacques Rousseau, Jeremy Bentham, and John Stuart Mill, treated group behavior as the result of individual agreement to a social contract. Arthur de Gobineau, a French aristocrat, gained wide attention for the mixture of individualistic and group behavior portrayed in his white-supremacist history of races and civilizations acting out their collective

[1] Bert Hölldobbler and Edward O. Wilson, *The Ants* (Cambridge, MA: Belknap Press of Harvard University Press, 1990).

destinies.[2] In the late twentieth century, rational choice theory arose to argue that group behavior can be explained as the combination of many individual preferences in various arenas of life. This individualistic outlook was accompanied by fierce critiques, among biologists and economists, of group-behavior theories; these critiques were countered by energetic defenses of the logic of group behavior.[3] A fuller focus on the history of group analysis in human affairs—for instance, in terms of behavior of families or civilizations—may eventually permit a reconstruction of the developing interplay of individual and group analysis in academic thought.

Only recently has there been a breakthrough in demonstrating that group behavior—among those who have chosen to act as a group—has clearly distinctive characteristics and cannot simply be reduced to individual behavior.[4] The scholarly world is heading toward a situation in which the frameworks of both individual and group behavior are recognized. The task for the future is thus for researchers to become clear on the distinctive characteristics of individual and group behavior for humans and to learn more of the interplay of humans acting in individual mode and group mode.

INDIVIDUAL BEHAVIOR

The dominant approach in biological and social disciplines has been to focus on individual behavior in itself, and to assume that group behavior can be explained in terms of individual actions of members of any group. Using individual behavior to explain group behavior is defined as a reductionist approach: such use of basic theory to explain complex phenomena has been applied in biological sciences and among the social sciences in sociology, economics, and psychology.[5] Within biology, the field of population genetics traces the alleles of genes (variations of a given gene) among members of a population. Populations (including sub-populations) are accumulations

---

[2] Arthur de Gobineau, trans. Adrian Collins, *The Inequality of Human Races* (Los Angeles: Noontide Press, 1966; first published 1854). Johan Gottfried von Herder had earlier sought to split the difference in developing a view that has recently been labeled as "holistic individualism." Vicki Spencer, "Towards an Ontology of Holistic Individualism: Herder's Theory of Identity, Culture, and Community," *History of European Ideas* 22, 3 (1996): 245–60.

[3] James S. Coleman, "Social Theory, Social Research, and a Theory of Action," *American Journal of Sociology* 91 (1986): 1309–35; George C. Williams, *Adaptation and Natural Selection: A Critique of Some Current Evolutionary Thought* (Princeton: Princeton University Press, 1966); Mancur Olson, *The Logic of Collective Action: Public Goods and the Theory of Groups* (Cambridge, MA: Harvard University Press, 1965); Vero Copner Wynne-Edwards, *Animal Dispersion in Relation to Social Behavior* (London: Oliver & Boyd, 1962); Richard C. Lewontin, Steven Rose, and Leon Kamin, *Not in Our Genes: Biology, Ideology, and Human Nature* (New York: Pantheon, 1984); Donald T. Campbell, "On the Conflicts Between Biological and Social Evolution and Between Psychology and Moral Tradition," *American Psychologist* 30 (1975): 1103–26.

[4] Tuomela, *Social Ontology*.

[5] For further discussion of reductionism, see Chapters 9–11.

of individuals. Population behavior is seen as arising from the dynamics and accumulations of individual behavior.[6]

The principal path for explaining group behavior through analysis of individuals has been based on the logic of *inclusive fitness*, arising out of population genetics. That is, the offspring of a parent share much of the same genome and thus have a special relationship in passing on their heritage. This analysis led to arguments that, under certain circumstances, interactions among individuals could lead to genetic expansion of traits of reciprocity, collaboration, and altruism that could affect whole populations.[7] An extension of this reasoning developed the theory of "dual inheritance" and the field of cultural evolution, in which a combination of brain-centered learning and gene-centered evolution brought greater levels of cooperation among groups of humans.

In the discipline of psychology, the field of social psychology is generally portrayed as the study of individual behavior in social context. Psychologist Donald T. Campbell, however, chose to link individual and group behavior. He coined the notion of "entitativity" to refer to the notion that individuals conceived of some social groups, such as family units, as having substantial reality.[8] In a project that gained more attention, Albert Bandura developed the notion of social learning, in which individual learning, stored in the brain, could be passed from generation to generation, eventually growing in social significance. This approach was adopted by ecologists Peter J. Richerson and Robert Boyd, who combined it with the logic of inclusive fitness to create a "dual heritage" explanation of changes at genetic and brain-record levels.[9]

For the social sciences more generally, the heritage of "methodological individualism" has been traced back more than two centuries by Lars Udehn. He traces this doctrine from the days of Hobbes and Locke to rational choice theory; he identifies three versions of methodological individualism that he calls social, institutional, and structural. He distinguishes those for whom individualism is an inescapable postulate (John Stuart Mill, Friedrich Hayek, and Karl Popper) from those for whom it is a heuristic device or research program (Weber and Schumpeter).[10]

---

[6] Robert Axelrod, *The Evolution of Cooperation* (New York: Basic Books, 1980); W. D. Hamilton, "The Genetical Evolution of Social Behaviour," *Journal of Theoretical Biology* 7 (1964): 1–52; John Maynard Smith, *Evolution and the Theory of Games* (Cambridge: Cambridge University Press, 1982); McElreath and Boyd, *Mathematical Models.*

[7] Hamilton, "Genetical Evolution." Biologist Lewontin and colleagues, in contrast, strongly supported the existence of a distinctive process of social evolution. Lewontin et al., *Not In Our Genes.*

[8] Donald T. Campbell, "Common Fate, Similarity, and Other Indices of the Status of Aggregates of Persons as Social Entities," *Behavioural Science* 3 (1958): 14–25.

[9] Albert Bandura, *Social Learning Theory* (Morristown, NJ: General Learning Press, 1971); Boyd and Richerson, *Culture and Evolutionary Process.*

[10] Lars Udehn, "The Changing Face of Methodological Individualism," *Annual Review of Sociology* 28 (2002): 479–507.

Individual game strategy. A well-known application of individual-level logic arises from game theory and takes the form of the Prisoner's Dilemma game. The game consists of two players. They are individual agents who pursue their own self-interest without the direction of an authority: each of them is unaware of and does not consider the totality of benefits to the two players. Payoffs are set and agents cannot communicate with or influence each other. In turn, each agent selects a move—either *cooperate* or *defect*—aware of the payoff they will receive. Irrespective of the opponent's choice, *defect* gives a higher payoff than *cooperate*. Yet both would be better off if each chooses to *cooperate*.[11]

Game: Prisoner's Dilemma			
		Agent 2	
		Cooperate	Defect
Agent 1	Cooperate	2.2	0.3
	Defect	3.0	1.1

Robert Axelrod gained wide attention for his concise 1980 book, *The Evolution of Cooperation*. In it, he argued that the most advantageous strategy in the Prisoner's Dilemma is the "tit-for-tat" strategy, in which a player first defects and thereafter matches the move of the other player. He further argued that cooperation based on reciprocity can grow from this base. Partly in response to Axelrod, game theory became popular for fun and profit in the financial world. Updated analyses have shown that Axelrod's preferred strategy was not exactly the most productive, yet the same results confirm the idea, exemplified in the Prisoner's Dilemma game, that cooperation based on reciprocity can grow over time.[12]

Topic 5.1: Individual behavior
Question: What is the best strategy for playing Prisoner's Dilemma?
Theory: Game theory
Evidence: results of numerous games of Prisoner's Dilemma in computer tournaments
Analysis category: verify a hypothesis
Answer to the Question: **Axelrod found "Tit-for tat" strategy to be most successful in large-scale trials**. Of the eight top algorithms, Tit-for-tat clearly scored highest.
Presentation: description of the computer tournament to which participants submitted algorithms that were applied in games of 200 moves.
Citation: Axelrod, *The Evolution of Cooperation*, 27–54.

---

[11] Axelrod, *Evolution of Cooperation*; McElreath and Boyd, *Mathematical Models*.
[12] Axelrod, *Evolution of Cooperation*.

Subsequent commentaries, while appreciative of Axelrod's analysis, have argued that there can be no definitive winner to the contest, as winners and losers in the game participate with a range of assumptions that is too wide to be fully accounted for.

Group Behavior

Different approaches to group behavior have arisen and declined in the arenas of both biological and social analysis. In the biological literature, there have been arguments for biological evolution in which groups of organisms are the unit of evolution; there have been arguments for systematic social change at the macro-level or phenotypical level of societies for which the mechanisms of change remained unspecified. Darwin himself considered the possibility that humans might have been able to undergo group biological evolution but backed away from the idea. Then in 1962, when biologist V. Wynne-Edwards sought to show that group-based biological evolution was feasible, the logic of population genetics was applied with resounding effect to argue that group-based evolution would almost always devolve into evolution of individual organisms.[13] Since then, the attempts to explain group behavior in biology have mostly focused on modeling groups through individual behavior.[14] Still, there continue to be arguments in support of the existence of group evolution, notably in viruses. A key test case has been that of the myxoma virus. In an attempt to eliminate the huge population of feral rabbits in Australia, the myxoma virus was introduced in 1950. This was a virus that had been isolated in Europe and was known to be lethal to rabbits. After its introduction through mosquitoes, the rabbit population fell by 99%, but then gradually grew. David Sloan Wilson and Elliott Sobert, among others, argued that the recovery of the rabbit population was a response to group evolution among the myxoma viruses, which were able to survive by becoming less virulent to their rabbit hosts. The debate has not ended.[15]

A major breakthrough in theorization of group behavior among conscious humans came in the field of philosophy with the work of John Searle and Raimo

---

[13] George C. Williams argued effectively that such "group selection"—selection of one group over others—was possible only under the most unusual of conditions. Wynne-Edwards, *Animal Dispersion*; Williams, *Adaptation and Natural Selection*, 97, 110–22.

[14] The discipline of biology must also address the task of making sense of group behavior among conscious humans, determining whether consciousness and communication make group behavior distinctive in humans as compared with group behavior in other species which do not have such advanced communications.

[15] Robert A. Wilson's 2004 review of the issue concludes that it is not resolved, though he is skeptical as to whether group evolution took place among myxoma. Robert A. Wilson, "Test Cases, Resolvability, and Group Selection: A Critical Examination of the Myxoma Case," *Philosophy of Science* 71 (2004): 380–401; David Sloan Wilson and Elliott Sober, "Reintroducing Group Selection into the Human Behavioral Sciences," *Behavioral and Brain Sciences* 17 (1994): 585–654.

Tuomela, also drawing on Michael Bacharach in economics. The breakthrough was the identification of "collective intentionality" as the additional factor that makes possible a distinctive sort of group behavior. In this view, group behavior is defined as collections of individuals and patterns of their behavior, which may differ according to whether the individuals recognize each other, respond to each other, share common intentions. Human group behavior relies on "collective intentionality," in which individuals agree and determine a joint intention (collectively), of which a we-intention is an individual dimension.[16] Tuomela, expanding upon Bacharach's work on group games, developed a demonstration of the impossibility of reducing group behavior to individual behavior. In this illustration of the principle through the Hi-Lo game, the two participating agents are members of a team. They communicate before but not during the game. Each agent selects a label, A or B, aware of the full payoff structure of the game; the agents are aware of their individual payoff and they are also alert to the total payoff for both players. To maximize the payoff, each player wants the result to be that they both choose the same label. Playing as a team, each agent can select the option that is best for the team.

Game: Hi–Lo				
			Agent 2	
			Hi	Lo
Agent 1		Hi	2, 2	0, 0
		Lo	0, 0	1, 1

This result is simply illustrated here. The two agents in the group each chose a level and they are rewarded based on the combination of their choices. The point is that the group strategy gives a unique choice (Hi Hi) where individual work yields irresolvable choices; and the group strategy reaches its conclusion in less steps. These points, suggested at the level of this simple game, can be confirmed in larger-scale analysis. For the full proof, see the analyses of Bacharach and Tuomela.[17]

Designing an institution. The notion of collective intentionality permits the identification of various types of human groups. Participation in group-level cooperation does not eliminate individual-level cooperation: the two interact with each other and with other factors. Groups may take the form of an I-group or a we-group. An I-group is a population in which each member acts on individual motivations and initiative—it is the same as the definition of a group within individual-level populations. A we-group is a

---

[16] Michael Bacharach, Natalie Gold, and Robert Sugen, eds. *Beyond Individual Choice: Teams and Frames in Game Theory* (Princeton: Princeton University Press, 2006); Norbert Elias, trans. Stephen Mennell and Grace Morrissey, *What Is Sociology?* (New York: Columbia University Press, 1978), 76–80; Tuomela, *Social Ontology*.

[17] Bacharach, *Beyond Individual Choice*; Tuomela, *Social Ontology*, 108–09, 201–03.

group of members unified and bounded by their collective intentionality, by which is meant their shared objective, recognition of their common interest, and agreement to act for the interest of a group. The alternative group situations thus include individuals in I-mode, who happen to share the same goals (implicit groups); individuals in we-mode, with collective commitment acting as a group (explicit groups); and groups that are dominantly organized by collective intentionality as we-groups, but within which there are individuals who form implicit groups in I-mode (complex groups).[18]

For the case of creating a new institution, a group of people sharing an objective agree to share the task of achieving it. You are asked to imagine a simple educational institution—a school—how it would be created and function, as well as what sort of groups it would include. As an extension of the exercise, you might also consider how the school would work with its clients, respond to difficulties, and reproduce itself after a generation.

Topic 5.2: Group behavior
Question: **To create a small private school for young children, how would you identify the members, goals, and beneficiaries of the new institution?**
Theory: Collective intentionality
Evidence: general knowledge on the operation of private schools in social context
Analysis category: propose a hypothetical relationship
Answer to the Question: (Thought experiment) **Identify members, goals, and beneficiaries of the school as an institution.**
Presentation: List the roles within the institution, identifying those who have agreed to be members of the institution, how participants function in institutional. Who are beneficiaries and what is their role?
Citation: Tuomela, *Social Ontology*, 34–36.

INTERACTION OF GROUPS AND INDIVIDUALS

Individuals and groups commonly overlap. Consider a we-group of individuals who are mostly pursuing common objectives. Some individuals within the group might choose to focus instead on their own individual objectives, performing their tasks somewhat differently. The result would therefore modify or diminish the effectiveness of the institution. If there were enough such independently minded individuals within a we-group, they could be seen as an I-group within the boundaries of the we-group. Details of these variations could be pursued—the point here is to emphasize the logical consequences

---

[18] For an early exploration of these ideas, see Peter Blau, "Introduction to the Transaction Edition," *Exchange and Power in Social Life* (New Brunswick, NJ: Transaction Publishers, 1986), vii–xvii.

that will arise from adding explicit theorization of group behavior to the common vague references to the existence of human groups. The types of mixes of groups and individuals are doubtless numerous and influential.

Distinguishing group and individual behavior. Individuals may be within a group yet act only in their own interest. Others support the group interest. How can you tell who is a committed member of a group? Is it what people say? Is it actions that they take? Is it types of expression or interaction with each other? For instance, if a person makes a critical comment, can you tell whether that person is acting as a member of a group? In this exercise, you are requested to identify some practical guidelines to identifying groups that are functioning collectively and individuals that are participating actively within the group.

Topic 5.3: Interactions: groups and individuals
Question: **Based on your experience, what criteria can show that several people are acting as a group?**
Theory: Collective intentionality
Evidence: observations of individuals in interaction
Analysis category: propose a hypothetical relationships.
Answer to the Question: (Thought experiment) **List 3 to 5 criteria that you find useful in confirming that several people are acting as a group**
Presentation: A list of criteria in order of significance
Citation: Tuomela, *Social Ontology*, 26–32, 70.

INDIVIDUALS IN NETWORKS

Network theory and graph theory have been developed significantly through representation of symmetric or asymmetric relations among discrete objects. Nodes and edges, labeled with names, are the key elements of these analytical networks, usually displayed in two dimensions. Paul McLean, exploring cultural issues through networks, has applied the logic of network analysis to networks of individual humans: such networks are constituted of individual human nodes and links among them.[19] Individuals enter the network space at birth and leave it at death or upon migration; they have links to family members and other individuals. Some individual nodes are more central than others; information can be passed through the network from node to node. In a restatement of McLean's formulation, I propose that such a network of individuals be labeled as an I-group or an "I-group-network." Specific issues to which this logic of individual behavior within networks can be applied are

---

[19] Paul McLean, *Culture in Networks* (Cambridge: Polity Press, 2017); see also Peter J. Carrington, John Scott, and Stanley Wasserman, *Models and Methods in Social Network Analysis* (Cambridge: Cambridge University Press, 2005).

the functioning of networks, individual behavior and social institutions, processes of learning, and individual emotions. Further questions can be posed about the characteristics of such I-group networks: Are the edges all alike or do they vary? What determines the boundaries of a network or the strength of ties within it?

McLean emphasizes both relationships and perspectives in study of networks. The ego-network, of whatever size, is viewed from the perspective of a focal actor. A dyad is any pair of nodes, viewed especially because of a specific tie between them. In triads, the relationship between any two nodes is affected by their relationships with the third node. With larger numbers of nodes, additional concepts can be introduced: cohesive subgroups, reachability, hierarchy, and centrality.

Topic 5.4: Networks of individuals
Question: What are differences among *ego-networks*, *dyads*, and *triads* in network interactions?
Theory: Network theory
Evidence: examples of nodes in various relationships, seen from various perspectives
Analysis category: define an ontology
Answer to the Question: **Ego-networks build out from a focal actor; triads enable relationships among two nodes to be affected by the existence of a third node.**
Presentation: Networks of multiple nodes are analyzed from perspectives ranging from *ego* to larger groups and with concepts defining their relationships.
Citation: McLean, *Culture in Networks*, 15–32.

GROUPS IN NETWORKS

As a next step in the logic of individuals, groups, and networks, I propose the identification of "we-group-networks": these are networks for which the nodes are we-groups. In a parallel to the case of individual behavior, the logic of group behavior through collective intentionality may be applied to the nature and functioning of social institutions, social networks, social evolutionary theory, and the role of emotions in social groups. A network with we-groups as nodes in which each of the we-group nodes pursues its own objectives would have characteristics and dynamics similar to those of an I-group-network. But in cases where we-group nodes that are participating in a network choose to express collective intentionality and a common goal, the network is an institution, a we-group network. For such a we-group-network, one can conduct the familiar analysis of the centrality of its various nodes.

Groups in networks. A further complication one can propose, in analysis of social networks, is the addition of a third dimension to the normally two-dimensional graphs. That is, within a we-group network, one may place the nodes in a hierarchy, with certain nodes being in ruling or highly influential roles; these hierarchies are operative within the boundaries of the network. The nodes are then linked by both horizontal and vertical edges. One could imagine a we-group of slaves under the domination of a we-group of masters. On the other hand, one could imagine the slaves as an I-group, under the dominance of the masters but not in agreement that the hierarchical relationship is legitimate. Other variations on these models show the potential of such social network analysis.

Religious organizations are well established; they emphasize both conformity and opportunities for independent expression. The Catholic Church, governed by the papacy and with priests as a separate class, is arguably very hierarchical. Sunni Islam has no priests and no papacy; the *imam* at the head of a congregation is a prayer-leader. Still, there are grand imams at major mosques and universities. The Pentecostal faith, beginning in 1900, has formally emphasized the immediate access of any believer to God, yet the church is divided into bishoprics headed by bishops serving under a presiding bishop. You are invited to explore online resources to see what indications you find on the organization and structure of these three major religious groupings.

Topic 5.5: Networks of groups
Question: **What are differences in the organization of groups within Catholic, Sunni Islamic, and Pentecostal religious faiths?**
Theory: Network theory; collective intentionality
Evidence: evidence on patterns of organization within major religious communities, available through internet searches
Analysis category: propose a hypothetical relationship
Answer to the Question: **Catholics appear most structured and Sunni Muslims appear least structured, in available information.**
Presentation: description of organizational structures within the three religious communities, comparisons of the structures, and conclusions. Comparisons should be attentive to the scale of comparison, from local communities to large-scale leadership.
Citation: Explore Wikipedia and other online resources to collect information and build your interpretation.

CHAPTER 6

Study of Human Institutions

In this chapter, I seek to show how the theory of social evolution—emphasizing social change through institutional evolution—provides a framework that encompasses and expands the knowledge in the social science disciplines. The approach focuses especially on the creation of key institutions, their replication, and the question of their social selection or de-selection. Among the key institutions are languages, units of production, and government—we study them through the disciplines of linguistics, economics, and politics.

Two definitions of institutions coexist within the world of social science. Especially among anthropologists but also among historians, institutions are seen as organizations. They are social structures created and maintained through the action of humans acting especially in groups but also as individuals—a lineage or a church is this sort of an institution. Among economists and many sociologists, however, institutions are seen as norms and ideas. In that view, institutions are seen as accepted *principles* that govern the behavior of humans—the notion of private ownership of property is this sort of an institution.

The anthropologists and historians, in adopting a group-based approach to institutions, have applied it broadly to human organizations across times and places in the human past and present, while economists and sociologists have restricted their focus on social norms to large-scale societies in recent centuries.[1] The focus in this chapter, therefore, is on institutions as

---

[1] Jonathan H. Turner, *Human Institutions: A Theory of Societal Evolution* (Lanham, MD: Rowman & Littlefield, 2003). My approach to institutions is distinct not only from the large-scale framing of Turner but also from the "old institutionalism" of John R. Commons and from the "new institutionalism" of Ronald Coase and Douglass C. North. As Udehn shows, the old institutionalism was descriptive and the new institutionalism took institutions as exogenous variables; both assumed all behavior to be individual and determined by social not biological factors. I analyze institutions as endogenous to the analysis, responding to their environment and to inherent structural dynamics, generating motivations and emotions in their personnel, and responding

organizations rather than as norms. This group-based approach to institutions has recently been strengthened greatly by the logic of collective intentionality, which has begun to clarify the distinctive dynamics of group behavior. The methods presented in this chapter introduce institutions created and revised in a wide range of human experience, to provide indications of underlying commonalities and unique characteristics in institutions.

INSTITUTIONS AS ORGANIZATIONS IN PROCESSES OF SOCIAL EVOLUTION

The approach here develops a group-based approach to institutions. It builds upon the theory of social change through institutional evolution that is articulated in Chapter 2, and on group behavior as discussed in Chapter 5. This section describes a general set of succeeding steps in the creation, functioning, and reproduction of social institutions.[2] The succeeding section applies this basic model of institutional evolution to several domains of society, showing the processes of institutional development in language, migration, exchange, governance, and other arenas of human life and experience.[3]

Definition and functioning of institutions. A social institution is an organizational form, a *we-group* with explicit or implicit objectives, involving human activities, behavior, and norms. Institutions are constructed, supported, and reproduced by members of one or more groups, requiring we-mode and accompanying outlooks in identity and activities. The institution defines ground rules on how to act for a collective item with a signified symbolic or social status in a collectivity. Specific institutions have dynamics

---

as well to the biologically based emotions in individual participants. Udehn, *Methodological Individualism*, 256–60. In addition, scholars working in the well-established tradition of physical anthropology in Japan have recently published a collection with an evolutionary approach to institutions among human and primates. The editors are cautious about defining "institution," since the usage of the term is quite different in Japanese than in English. Nevertheless, "If we consider that 'institution' is a 'situation' or 'thing' that externally influences a face-to-face interaction, a violent third party is a strong contender for 'institution'," Kaori Kawai, ed., *Institutions: The Evolution of Human Sociality*, trans. Minako Sato (Kyoto University Press, 2017), 10.

[2] Searle and especially Tuomela, in defining institutions in terms of we-groups, opened the door for rethinking of institutions in more tangible terms. On the three requirements of collective intentionality in the formation of a social group, see Chapter 3.

[3] In another work, I have given explicit discussion to a number of major institutions, as follows. For the Pleistocene era: syntactic speech, community, custom, migration, religion, marriage, we-group network, workshops, ceramic, and confederation. For the early and mid-Holocene era: society, agriculture, animal husbandry, judiciary, slavery, states, metallurgy, commercial institutions, systems of transportation, coinage, literacy, schooling, military force, water supply. For the late Holocene era: large-scale religion, empire, colonies, university, scientific knowledge, oceanic network, joint-stock companies, capitalist institutions, insurance, and the republic of letters. For the Anthropocene era: national states, factory production, trade unions, corporations, labor-intensive industrialization, and international organizations. Manning, *History of Humanity*.

arising from the character of their activities, so that institutions of agriculture experience choices that are different from the choices facing institutions of warfare.[4] More specifically, the general process of creating and preserving a social institution requires the following steps, all carried out by a relevant we-group or set of we-groups:

- *Create the institution.* Members join and agree on the institutional objective and function.
- *Identify institutional beneficiaries.* Those expected to benefit from the institution should be individuals and groups within the institution and, commonly, beyond the institution.
- *Identify a "generation."* A generation is the time period within which the institution must be reproduced if it is to survive.
- *Perform the institutional function.* The institution functions through collaborative work or division of labor. Its benefits are expected to reach the beneficiaries.
- *Respond to institutional dynamics.* Operation of a new institution necessarily reveals unsuspected dynamics that are specific to the institutional domain. The institution must adapt to these dynamics.
- *Construct and update an archive.* Information needed for reproducing the institution for the next generation must be stored and available for recall.
- *Assess fitness of institution.* The institution will be assessed for its fitness by its members, beneficiaries, and other influences, roughly every generation.
- *Reproduce and regulate the institution.* Members reproduce the institution, drawing on the archive for direction. Regulation of the institution can take place from within and without, dependent on its apparent strengths and weaknesses.

Institutions in practice. Institutions, as described above, can sometimes be portrayed more clearly when they are interpreted as networks, thus making explicit their spatial dimensions. Thus, a we-network is a social institution like any other we-group except that it is also defined in spatial terms. Analysis of such a network involves accounting for the appearance, in many social situations, of individuals, I-groups, and we-groups. This enables distinctions such as that between a network of communities linked by migration and exchange of goods (comprising an I-group-network) and a network of communities linked in a compact for common defense against invaders (comprising a

---

[4] Tuomela, *Social Ontology*; Manning, *History of Humanity*. Along with the analysis of institutions, I seek to draw inspiration from the work of Robin Dunbar, to hypothesize the approximate size of human groups associated with various social structures over time. See, for instance, Robin I. M. Dunbar and Richard Sosis, "Optimising Human Community Sizes," *Evolution and Human Behavior* 39 (2018): 106–11.

we-group-network). Further, treating hierarchies as vertical networks has the advantage of sustaining an analysis that keeps track both of hierarchy and of networks across horizontal space, as well as interactions of the two. Finally, to get back to the contending definitions of institutions—focusing, respectively, on organizations and norms—the place of norms within institutions becomes clearer when we-groups, networks, and hierarchies show the basis for study of norms: their creation, reproduction, and impact.

The role of representation in social institutions. The details of activity within institutions call for representation—innovative thinking. "Representation" is defined here as the modeling, formalization, and interpretation that is created in response to an aspect of the world as it is or might be. At one level, representation takes place within the mind of an individual—internal and conceptual formulation in the individual mind to represent a thing, a process, or an idea. At another level, the exchange of representations generates discourse, debate, and innovation—especially through speech within a we-group context. Observation of group-level culture and its practice inspires responses from individuals and groups. Representation is commonly the modeling and interpreting of one aspect of the world for expression in another arena: representations thus cross-boundaries of material culture and expressive representation in visual art, dress, dance, or song, with new media for representation developed steadily through the course of human experience. Representation or conceptualization is required at every stage of human life, but especially at the stages of conceptualizing an institution and at the stage of regulating it.

HISTORICAL INSTITUTIONS

Major categories of institutions are briefly introduced here: the order of presentation is from early institutions to some recently developed institutions. Each set of institutions took form especially in response to the domain and task for which it was constructed. The dynamics of each institution were given by the inherent characteristics of that domain—they can be studied through one or more social science disciplines. Each example gives particular attention to a historical method and its application to the topic of that institution.

Language. The questions raised in the 1950s by Noam Chomsky are now receiving detailed study and interesting answers. Distinctions are now being clarified among communication, capacity for language, the logic for formulating sentence-level propositions, and syntactically spoken language. It is my view that any theory of human evolution must include a specific argument on when and how syntactical language came to be used.[5]

---

[5] In addition to Chomsky's work on the logic underlying language and Greenberg's work on language classification, Greenberg showed the importance is language universals—the similarities in language structures and patterns that are being identified. It is not yet possible to suggest

The institutional practice of language was presumably created by people who found a way to share conscious communication among members of a community. The function of the institution was to develop language that would allow communication by speech in all social situations. The beneficiaries included all who spoke a language, and some who spoke none. The dynamics of speech included the steady expansion in vocabulary, the variations in pronunciation and the shifts in syntax with time. In addition, words were "borrowed" from one language to another in the course of cross-community communication and sharing. The archive of language, at base, was in the spoken practice of young and old; the generational time for renewal was much the same as the demographic generation between parents and children. The issue of fitness was: How useful was the institution of language, as it had been practiced in the current generation, to the speakers who were members of the institution, to any other intended beneficiaries of language, and to the community overall? If the response was generally positive, one could continue to reproduce the speech community, with the possibility of regulation (by those who sought change).

Borrowing of words. For the study of language and the historical reconstruction of its institutions, one must rely primarily on the discipline of linguistics. Much of linguistics focuses on the study of changes, usually in the short term, in syntax, lexicon, and phonology. The subfield of historical linguistics focuses on longer-term linguistic change and the evolution of language families.[6]

Christopher Ehret, in the selection cited here, has shown the range of patterns in "borrowings," as words move from one language to another. With examples of borrowings from Spanish to English, from Algonkian to English, and East African cases, Ehret builds up to a chart showing that large-scale borrowings (as from French to English) require two to three centuries, while small-scale borrowings of technical terms ("telephone") happen very rapidly. For basic terms such as small numbers and body parts, they are rarely borrowed.

---

implications of language universals for human history, but the developing sense of the coherence of language overall helps to reaffirm the distinctive nature of syntactic language as a system unlike any other. Joseph Greenberg was as central to development of the literature on language universals as he was on language classifications. He remained convinced that all languages could in principle be traced back to that original language. See William Croft, *Joseph Harold Greenberg, 1915–2001: A Biographical Memoir* (Washington, DC: National Academy of Sciences, 2007), 23; Comrie, *Language Universals*, 23–27; and Greenberg, et al., *Universals of Human Language*, 4 vols.

[6]Topic 1.1, on the formation of a language community, opens discussion on the dynamics of language communities. For a discussion and maps on 14 language phyla and the migration of speaking populations throughout the world, see Patrick Manning, "Language Resources," www.cambridge.org/humanity.

> **Topic 6.1: Linguistics**
> *Question:* **How long does it take for words to be borrowed across languages?**
> *Theory:* Historical linguistics
> *Evidence:* cases of borrowing in North America and East Africa
> *Analysis category:* estimate parameters (in relative frequency)
> *Answer to the Question:* **some single words are borrowed within days; general borrowing takes one to three centuries; basic vocabulary is rarely borrowed.**
> *Presentation:* discussion with examples for single-word, restricted, and general borrowing, with a table to comparing each pattern by type of vocabulary and how transmission took place.
> *Citation:* Ehret, *History and the Testimony of Language*, 82–102.

Migration. Human migration takes place through two main processes: *general migration*, diffusion of population to extend its habitat or occupy a new habitat; and *cross-community migration*, to link existing communities.[7] The objective and function of the institution of migration are to facilitate human mobility, especially to maintain contact with spaces and communities. The structure of the migratory institution centers on those who send and receive migrants and on the migrants themselves. The institutional practice of migration focuses on the tasks of facilitating migration outward from and into the home community. In it, one must identify migrants, their directions and tasks; dispatch and receive immigrants, house them, and assign tasks to them. Where migration is common, additional institutions form to facilitate or benefit from the movement of migrants. Further, one must identify the beneficiaries of migration and distribute benefits among them. Institutions of migration must respond to the inherent dynamics of migration, such as mortality in movement, other losses, and the need for migrants to learn languages and new cultural skills. An archive of migration practices must be built up in practices of those who govern migration; it must prepare to renew the migratory system. The fitness of migratory institutions must be assessed—for instance, to determine whether the number and direction of migrants is sufficient—after a generation that is roughly the same as a demographic generation. The reproduction of the migratory institution (and its archive) is also carried out especially at the end of a generation.

---

[7] On general migration, see Hugh Dingle, *Migration: The Biology of Life on the Move* (Oxford: Oxford University Press, 1996); on cross-community migration, see Patrick Manning with Tiffany Trimmer, *Migration in World History*, 3rd ed. (London: Routledge, 2020). I propose here to use the term "general migration" as a synonym for what I have previously called "colonization."

Human migration relies on certain basic impulses, yet it has changed in character along with social transformations. Humans have repopulated the Earth in characteristic fashions as social, ecological, and economic conditions change.[8] Migration fills in new gaps in habitat as they develop with changes in ecology and technology. Disciplines that are important in documenting migration over the long term include historical linguistics, genetics, archaeology, and written records.[9] A fascinating source of information on early migration comes from analysis of the bones and teeth of individuals from archaeological sites. Analysis of strontium isotopes in those bones and teeth can indicate of where the individuals were born or lived earlier, by linking their isotope ratios to those known from worldwide mapping.[10] Migrations in recent centuries are documented especially through written records, analyzed through demography and sociology.[11] The analysis of migration considers not only the movement of people but the social and cultural changes that result from migration.

Institutions of migration. Theories of migration, especially for well-documented migrations of recent times, show how migration is both facilitated and restricted by social institutions. The recent survey of migration theories cited here gives numerous examples of institutions that function in migration. These include families of origin, communities of reception, diaspora communities, and organizations transporting migrants. Ports and borders are crucial bridges or walls. Formal organizations include governments that encourage, inspect, limit, document, or forbid migration; relief organizations, organizations providing lodging, food, and clothing during travel; plus banks and money lenders. There are employers in agriculture, mines, industry, domestic employment, and cultural work. The interplay and functioning of these institutions do much to define the flows of migrants and the social roles they play.

---

[8] For methods and additional data on projecting migration from language distribution, see Manning, "Language Resources"; see also Patrick Manning, "Homo Sapiens Populates the Earth: A Provisional Synthesis, Privileging Linguistic Data," *Journal of World History* 17 (2006): 157-58.

[9] For earlier efforts comparing and linking language and genetic data, see Luigi Luca Cavalli-Sforza, et al., "Reconstruction of Human Evolution: Bringing Together Genetic, Archaeological, and Linguistic Data," *PNAS* 85 (1988): 6002–6006; and Cavalli-Sforza, *Genes, Peoples, and Language* (New York: North Point Press, 2000).

[10] Shomarka Keita, "A Brief Introduction to a Geochemical Method Used in Assessing Migration in Biological Anthropology," in Jan Lucassen, et al., eds., *Migration History in World History* (Leiden: Brill, 2010), 59–74. Strontium closely resembles calcium, which is why it is visibly present in bones and teeth.

[11] Dirk Hoerder, *Cultures in Contact: World Migrations in the Second Millennium* (Durham, NC: Duke University Press, 2002).

> **Topic 6.2: Migration**
> *Question:* **What institutions are highlighted in migration theory?**
> *Theory:* Migration theory
> *Evidence:* social institutions related to migration, as described in theories of migration
> *Analysis category:* define an ontology
> *Answer to the Question:* **Major institutions include family, border, transportation firms, host community.** Many more can be listed and their functions can be shown.
> *Presentation:* Identify social institutions discussed in migration theory; organize them into an ontology reflecting their roles and relationships
> *Citation:* Manning, *Migration in World History*, Appendix: Migration theory and debates, 222–245.

Exchange and Commerce. The objectives and functions of exchange are for a community to obtain valuable material culture, perhaps in exchange for existing resources. The tasks are to identify desired resources and their location, travel, collect directly or through exchange, and return home. Simple exchange includes implicit collaboration among individuals who make exchanges, without committing themselves to larger groups. The dynamics of an exchange network include losses in travel, assessing value of goods, and the problems of transport. Knowledge must be stored on goods, routes, centers of demand and supply. A generation for an exchange network is as short as the length of a voyage; as long as the life of a merchant. With formal commerce, institutions based on we-networks are created for elements of exchange: ports, markets, and transport. In these cases, one must reproduce the institution each generation, assessing the fitness of the institution and its ventures. Regulation of the institution depends as well on its beneficiaries, who may be traders, consumers, bankers, or thieves.

David Hancock found that economic historians had focused on changing productivity as the explanation of economic change up to 1940; thereafter, social historians focused on the choices of consumers as the main stimulus to expanding Atlantic trade. Hancock used the case of trade in Madeira wine to combine the two interpretations. Here, Hancock's interpretation is rephrased to show the balance of institutions and informal networks in the success of Madeira wine.

In a historical approach, Hancock explores over a century in production and trade in Madeira wine, showing how numerous changes in production and quality were facilitated by flows of information as well as links among institutions. The participants and agents in his story include consumers of wine—in England, North America, and the Caribbean—scattered but thirsty, and expressing their tastes to the distributors. Producers of wine in Madeira, shippers of wine on the Atlantic, and distributors in ports around the ocean—these were all firms, institutions seeking a profit. Governments and their regulations were also institutions. One more institution was the cartel formed by producers in Madeira to keep prices high and keep competitors out, in the mid-nineteenth century.

The dynamics of the story resulted mostly from consumer expressions of their tastes. First consumers liked their Madeira fortified, so brandy was added. Second, consumers liked the fortified wine better when it was shaken and homogenized. Third, consumers liked aged Madeira better. Fourth, when the wine was heated it became preferable. First it was the shippers who benefited from shaking the wine and heating it; in each case, the producers learned how to handle that task themselves. The various firms competed as one expects; but it was the information, passing through informal networks from consumers to all participants, that enabled the overall growth of Madeira as a luxury product.

Topic 6.3: Exchange & commerce
Question: **In the transatlantic commercial system of the eighteenth century, identify aspects of it that were institutions and aspects that were networks.**
Theory: economics; sociology; history
Evidence: Hancock's description of production and consumption of Madeira wine
Analysis category: define an ontology
Answer to the Question: **for Madeira wine, communication networks linked individual consumers and institutions of producers, shippers, and distributors, thereby transforming the produce and the industry.**
Presentation: Narrative of institutions and networks in wine trade
Citation: David Hancock, "Commerce and Conversation in the Eighteenth-Century Atlantic: The Invention of Madeira Wine," *Journal of Interdisciplinary History* 29 (1998): 197–219.

Institutions of leadership. Leadership has been about more than governance—it includes command, persuasion, invention, facilitation, critique, coordination, and justice. All these types of leadership have been required by societies encountering new situations, but not all of them at once. Kent Flannery and Joyce Marcus have given close attention to one major choice in leadership of Holocene-era societies: they call it the difference between achievement-based and hierarchy-based leadership.[12] In the first case, outstanding individuals gained prestige and recognition for their individual accomplishments; their successors must meet the same criteria. In the second

---

[12] Kent Flannery and Joyce Marcus, *The Creation of Inequality: How Our Prehistoric Ancestors Set the Stage for Monarchy, Slavery, and Empire* (Cambridge, MA: Harvard University Press, 2012), 547–57.

case, leaders gained influence and also the ability to pass their role on to a chosen person of the next generation, typically a son.

The institutions of the two types of leadership were quite different. Achievement-based leaders were called upon to distribute their wealth to the community in ceremonies near the end of their life. Hierarchy grew especially as well-off families gave loans or took in the poor from other families, so that debt slavery grew into permanent hierarchy. It may be that achievement-based leadership was most effective in building public works, while hierarchical leadership was best in war. Replacement of leaders was a time for beneficiaries of the system to call for reform and regulation.

Topic 6.4: Institutions of leadership
Question: **What are the relative strengths of achievement-based and hierarchy-based leadership?**
Theory: social anthropology
Evidence: anthropological narratives of social change in the Holocene era
Analysis category: propose a hypothetical relationship
Answer to the Question: **Communities preferred achievement-based leadership yet hierarchy-based leadership eventually came to dominate.**
Presentation: Discussion of numerous examples shows that most communities preferred to support achievement-based leadership. Nevertheless, devices such a debt slavery enabled hierarchy-based leadership to grow increasingly.
Citation: Flannery and Marcus, *The Creation of Inequality*, 547–57.

Literacy. The objective of the institution of literacy is to record, preserve, and exchange information on selected topics. The institutional practice of literacy centered on the tasks of creating and preserving a writing system and training specialists to use it. Researchers have focused more on the scripts themselves than on the institutions supporting literacy. Materials may exist for describing the training and collaboration of scribes, the control of literacy and other dynamics, the archive for preserving literacy, and questions of who benefited from literacy. The principal invitations of literacy were in Egypt (ca. 3400 BCE), Mesopotamia (ca. 3200 BCE), China (ca. 1200 BCE), and Mesoamerica (ca. 250 BCE). All of these began as logographic scripts—images—and gradually developed sounds. Specialist scribes learned and preserved the written word.

Later came multiple revisions, makeovers, and diffusion of writing system. The most crucial makeover was the invention of Semitic writing systems in the Egyptian desert as early as 1850 BCE, perhaps by speakers of Canaanite language. Albertine Gaur recounts the rediscovery of this invention of the first Semitic script, modeled on Egyptian hieroglyphics but skillfully simplified. Roughly 25 characters were needed, each with a consonant sound. It could be learned much more easily. With small changes, Phoenician emerged by 1000 BCE and spread across the Mediterranean, giving rise to the Greek and Latin systems; Aramaic emerged immediately after and spread eastward to Persia, yielding the writing systems of Central, South, and Southeast Asia.

> **Topic 6.5: Literacy**
> *Question:* **How did literacy spread?**
> *Theory:* institutional evolution
> *Evidence:* written inscriptions in many languages
> *Analysis category:* propose a hypothetical relationship
> *Answer to the Question:* **A few inventions of literacy; preservation; lots of remakes of writing systems.**
> *Presentation:* The presentation is a succession of narratives of the creation of individual scripts
> *Citation:* Gaur, *A History of Writing*, 88–92.

Capitalism. Too big and complex to be a single institution, capitalism is a collection of institutions. Overall, its objective and practice are to support production and consumption of goods and services for local and distant trade to yield profit. Dynamics of capitalism include business cycles, labor unrest, monetary systems, taxation, and war. The many institutions of capitalism have their own archives and processes of reproduction, with potential regulation by those who benefit most. Is there a way to break down this massive and diverse system to issues of manageable size? It is explored through political, social, and economic theories.

Even in the seventeenth-century rise of capitalism, many of its key institutions had long existed: profit-making business firms, money and banking, marketplaces, transportation systems. I argue that the keys to emergence of capitalism were new networks of capitalist leaders, able to influence governments to follow pro-capitalist policies in taxation, regulation, and warfare.[13] By the late nineteenth century, capitalism had developed industrial production, international banks, expanded empires, and public education systems adding skills to workers.

In the twentieth century, capitalism faced the rise of socialism and certain tendencies in nationalist movements to maintain independence from the flows of international trade. After World War II, a powerful war machine kept pressure on socialist states and independent-minded nations. The United Nations and its institutions coordinated relations among capitalist powers. International institutions (IMF, World Bank) coordinated money flows and investment in development. With decolonization, new structures needed to

---

[13] For an institutional analysis of early days of capitalism, see Manning, *History of Humanity*, 178–91.

be set up to ensure that ex-colonies remained within the capitalist orbit: the establishment of OPEC is an important example. At the end of the century, major transformations expanded the capitalist system: the fall of the Soviet Union, of Eastern European socialist states, and the shift of China and Vietnam from socialist to capitalist economic organization. One essential institution for capitalism, often escaping notice, is the courts and the legal profession. While systems of law are important in regulating daily relations among common people, the high-powered work in law and justice is adjustment of regulations to favor the interests of powerful corporations.[14]

> **Topic 6.6: Capitalism**
> *Question:* **What new institutions enabled capitalism to expand in the twentieth century?**
> *Theory:* institutional evolution
> *Evidence:* descriptions of institutions that sustain capitalism
> *Analysis category:* propose a hypothetical relationship
> *Answer to the Question:* **New institutions governed money, trade relations, response to decolonization, military pressure on socialist states, reliance on courts and law.**
> *Presentation:* Trace major changes in character and extent of capitalism since 1900; hypothesize institutions likely to have facilitated those changes.
> *Citation:* Milanovic, *Capitalism, Alone*, 1–11.

---

[14] Branko Milanovic, *Capitalism, Alone: The Future of the System That Rules the World* (Cambridge, MA: Belknap Press of Harvard University Press, 2019).

CHAPTER 7

Emotions and Human Nature

The issues of emotions and human nature, widely discussed at a speculative level, have either been left entirely out of analyses of human evolution or have been discussed very inconsistently. New research appears to be changing this situation. Because of the advances in the disciplines of biology, psychology, and their overlap, there are now analytically sophisticated and experimentally detailed analyses of several aspects of emotions. The results of this work are intriguing though not yet definitive, and they suggest some important debates to be resolved. Further, the new work on emotions stops short of giving firm implications for the understanding of human nature, but it does suggest that the deep and vague issue of human nature may soon be worth exploring in the context of human evolution.

EMOTIONS IN INDIVIDUALS

What are the processes linking genetic emotions to their expression, to the subjective recognition of the emotions, and to reports on them? At the phenotypical level, there are numerous studies of individual emotions or behavior, especially violence and cooperation. These center on efforts to determine whether the behavior is inherent and unchanging, or whether it changes with social circumstances. Theorists of genetic emotions assume an emotional state in the context of stimuli, yet assume these states can be influenced by individual volition.

Emotions are functional mental states in individuals that are typically caused by sensory inputs, that typically cause behavioral outputs, and that also cause changes, and can be caused by, other mental states like perceptions, memories, attention, and so forth. Topics within the study of emotions include lists of basic emotions, such as happiness, sadness, fear, anger, surprise, and disgust. Yet it is commonly unclear whether they are defined at the phenotypical or genotypical levels.

The model of emotions advanced by Ralph Adolphs and David J. Anderson, working within the field of ethology, is a comprehensive model at the individual level, beginning with a functional definition of emotions. In it, emotions are defined by what they do (by their causal inputs and outputs), rather than by how they are instantiated in the brain in any specific case (which may vary from species to species, and even individual to individual).[1] As Adolphs and Anderson emphasize, emotions can be studied without needing to study subjective feelings.

> **Topic 7.1: Identifying emotion in animals and humans**
> *Question:* **How are feelings related to emotions?**
> *Theory:* Emotions in biological context
> *Evidence:* stimuli influencing emotional states, behaviors responding to emotional states
> *Analysis category:* propose a hypothetical relationship
> *Answer to the Question:* **Feelings are conscious experiences of emotion; emotions are biological states in all animals, inherited and implemented by neural mechanisms.**
> *Presentation:* concluding chapter of a book, summarizing earlier details
> *Citation:* Adolphs and Anderson, *Neuroscience*, 308–13; see also 313–26.

Emotions in Groups

Lisa Feldman Barrett's approach to analyzing emotions focuses on speaking humans. It gives attention to the initial learning of human infants, emphasizing the different sorts of learning they do. Infants are especially interested in sounds made by other humans. In their learning practice, known as *statistical learning*, infants seek to categorize patterns and learn to dissect the streams of sounds they hear into words. They focus on what becomes their native language, the sounds they hear from those around them. When others use a word to identify a concept, the child uses the word to label a reference. In this way, a child can learn the concept of "chair" but also the concept of "angry," by relying on others for the concept and by populating the concept in their own mind with the furniture or the behavior labeled according to the term. Second, with regard to facial expressions, Barrett argues that, empirically, neither infants nor adults are able to categorize them effectively. (Barrett thus contests the notion that faces and facial expressions provide a "fingerprint" according to which each individual can be recognized—she thus joins the many who are skeptical of current investment in facial-recognition software.)

---

[1] Ralph Adolphs and David J. Anderson, *The Neuroscience of Emotion: A New Synthesis* (Princeton: Princeton University Press, 2018), 41.

Children's concepts of emotions come therefore from their experiences with other people; emotions are labeled by words the children have learned. The discourse on emotions in which children participate will be updated with age and experience, but the discourse passes through the concepts in the mind of each. Underlying emotional states are still there: they are influenced by stimuli and they generate behavior. Further, emotional states generate "feelings" as part of their output. As Barrett puts it, the feelings are universal but the expressed emotions of fear and anger are not.[2]

Barrett's approach does not go as far as explicit discussion of group behavior. Yet she describes individual behavior and interactions of individuals, so that one could assemble several such individuals into the model of a group. She implies that group members, in their discussions and interactions, respond not to the direct emotions or even feelings of others, but to the combination of their own understanding of emotions and the understandings of others.

Topic 7.2: Exchange of emotions in a community
Question: **How do humans construct their emotions?**
Theory: theory of constructed emotions
Evidence: observations of children learning about emotions
Analysis category: propose a hypothetical relationship
Answer to the Question: **Children learn categories of emotions in experience with adults in which words are associated with actions or items**. Children continue learning with time, developing their ideas of emotions.
Presentation: summary of general argument with individual examples
Citation: Barrett, *How Emotions Are Made*, 94–103.

HUMAN NATURE: BIOLOGICAL ASPECTS

The term "human nature" is widely and variously used. While emotions refer to specific feelings, human nature refers to a general summary of human behavior. Human nature is present in the title of many books and articles but only rarely does it seem to be a specific object of research. In the work of Joseph Lopreato, "human nature" refers to patterns of phenotypical or directly observable human characteristics and human behavior. Human nature, while arguably observable in its expression, is widely understood to be based on underlying biological makeup. Researchers into human nature

---

[2] Referring to feelings of pleasure and displeasure, as with regard to the taste of walnuts, Barrett argues that "These feelings are universal, even as emotions like happiness and anger are not, and they flow like a current through every waking moment of your life." Barrett, *How Emotions Are Made*, 56.

assume it is a multi-scale phenomenon, ranging from the genetic level to the level of behavior and expression, but with no consensus on the details of the scales and the mechanisms at each scale. Emotions are seen as an important constituent of human nature.

Historian Carl Degler traces the gradual reaffirmation of a biological base to human nature up to the 1980s. E. O. Wilson showed the place of human nature in his reductionist sociobiology project, in which genetic explanation provides the full explanation for all human social behavior. Steven Pinker attacked the notion, which he attributed to John Locke, that all human behavior stems from nurture rather than nature. Lopreato sought to show both biological and social dimensions of human nature.[3]

Topic 7.3: Human nature in biological terms
Question: **Does there exist a biologically based human nature?**
Theory: biology, psychology
Evidence: observations of human behavior
Analysis category: propose a hypothetical relationship
Answer to the Question: **Human nature, though not yet clearly defined, must have a substantial biological component**
Presentation: various arguments in support of biologically-based, individual human nature, yet with differences among them.
Citation: Degler, *In Search of Human Nature1*, 59–104.

HUMAN NATURE: SOCIAL ASPECTS

Lisa Feldman Barrett offers what she calls a "constructivist worldview." She argues that, because the brain is cut off from the world, it provides organisms with an "affective reality"—one experiences what one believes. The human brain, because of its size and its multiple connections, allows for "degeneracy," in which the functions of individual brains can be performed in many ways. As a result, humans have one kind of brain with a common set of networks, yet quite different types of mind emerge during each individual experience of life. Barrett proposes a routine through which people advance their understanding of the world. To begin at an arbitrary point, they begin with affective reality, use their curiosity to test it, then use the concepts they have learned to model the outside world. Then they encounter social reality, gradually learning from it. In the course of this recurring

---

[3] Carl N. Degler, *In Search of Human Nature: The Decline and Revival of Darwinism in American Social Thought* (New York: Oxford University Press, 1991); Joseph Lopreato, *Human Nature and Biocultural Evolution* (Boston: Allen & Unwin, 1984); Edward O. Wilson, *On Human Nature* (Cambridge, MA: Harvard University Press, 1978); Steven Pinker, *The Blank Slate: The Modern Denial of Human* Nature (New York: Viking, 2002).

routine, there are repeated interactions of perceiver-dependent concepts with perceiver-independent reality.

Can one link this logic to a sort of human nature at the social level? It is my hypothesis that biological emotions exist (with their typology to be set by the Adolphs-Anderson process). In the course of the expansion of hominid brains, there was modification of emotional capabilities. That is, cultural evolution developed learning but also attitudes; behavioral predispositions arose at the phenotypical level; and expression of emotions developed with experience but especially as expressed through speech. Later in human history, changes in social conditions—especially the functioning of institutions—encouraged various specific types of behavior. As a result, a *zeitgeist* might result as a common expression of outlook and ideology linked to certain institutions and to behavior and emotions that are central to those institutions. Multiple *zeitgeisten* might exist, sustained by different sets of institutions. As a result, human nature becomes more group-based as institutions expand.

Topic 7.4: Human nature in social terms
Question: **Is there a human nature specific to group behavior?**
Theory: constructivist psychology
Evidence: observations on expressions of emotions in individuals and groups
Analysis category: propose a hypothetical relationship
Answer to the Question: **If individual minds differ and change, groups of individuals must have behavior specific to the group – thus, there exist many variants in human nature, by group**.
Presentation: assembly of a speculative hypothesis about group behavior, developed out of the specifics of Barrett's analysis of individuals
Citation: Barrett, *How Emotions Are Made*, 278–90.

What are the ways of exploring the significance of emotions in human dynamics and in human evolution? What are the factors determining human nature and its changes? While we don't have ready answers, it is arguably important to advance and test hypotheses in order to expand our knowledge. I conceive of all the above theories and methods as analytical tools that can be put to work on the project of locating missing links that separate our understanding of biological evolution and processes of social change. Clearly, the two must function on different scales. Yet there may be processes that link biological evolution and social change, directly or indirectly. I see the greatest promise for locating missing links in the possibility that the emotions of individual humans, while biologically conditioned, are tied to the emotions of groups of individuals. Such links would operate in the setting of social institutions that themselves are governed by certain inherent dynamics. Biological emotions of individuals, mediated by their socialization as infants, would be mediated again to become emotional outlooks shared among those associated with a given institution. This array of issues suggests to me that we might not be far from tracing interactions up and down the evolutionary scale—thus getting beyond the discourse of the past two centuries, in which biological change and social change are treated as completely separate processes.

PART II

Disciplines and Theories

CHAPTER 8

Disciplines and Their Evolution

Of the theories and historical methods summarized in the chapters of Part I, most were developed in the years after 1850. The academic framework underlying this new knowledge has been that of the scholarly *discipline*. While academic disciplines came to be recognized gradually, one may identify the year 1850 as marking the serious launch of disciplinary tradition in academic life, based on the German model of universities. A discipline was a community of scholars devoted to study of a specific academic subject. The scholars in each discipline shared a tradition of inquiry, an intellectual paradigm, and a common agenda for research and teaching. A discipline was generally led by university professors employed in academic departments within research universities. As research universities gained strength, teaching colleges extended their model.[1] From the mid-nineteenth century, a disciplinary map had become clear, although it underwent revision and updates with the recurring expansion in knowledge.

The chapters of Part II provide a narrative of disciplines and disciplinary change from 1850 to the present. Within this narrative, I give principal attention to individual disciplines and to the grouping of disciplines into physical, biological, and social sciences, with brief references to the arts and humanities. The expansion of knowledge brought an increasingly specific character to each discipline over more than a century of disciplinary specialization. In this era, the disciplines relied especially on their place in expanding universities but also worked within government research institutes and private corporations; they relied further on disciplinary associations and publications. Within the disciplines, I give priority to the disciplines grouped into biological and social sciences. Through this focus, I emphasize the gradual clarification of a discourse on "human evolution"—how overall evolution of human existence has relied on interacting processes of biological, cultural, and social evolution.

[1] Anthony Biglan, "The Characteristics of Subject Matter in Different Academic Areas," *Journal of Applied Psychology* 57 (1973): 195–203.

A third and broader objective in Part II is to trace modifications in the structure of human knowledge since 1850. While this narrative focuses especially on the strengthening of disciplinary structures, it also gives attention to two other areas of significant change: the growing recognition of multiple scales of reality and the increasing interest in exchange of knowledge across disciplinary boundaries. Beyond academic knowledge, accompanying changes in human belief and knowledge during the same era have been the rise and decline in racial discrimination and segregation, as well as the rise and decline of socialism in ideology and governmental systems. In practical terms, these chapters provide background on the wide range of methods for human history that have been introduced in Part I.

Scholarship Before Disciplinary Structures

The best-known achievements in expanding knowledge from the seventeenth to the early nineteenth centuries emerged from researchers whose work ranged across numerous fields of study. Most were self-supporting or relied on contributions from wealthy patrons. Among them were Giambattista Vico, Isaac Newton, Benjamin Franklin, George-Louis Leclerc (comte de Buffon), Johann Wolfgang von Goethe, and Immanuel Kant—a remarkable list of polymaths. Rare indeed were scholars such as Carl Linnaeus, who managed to devote a lifetime to a subject as consistent as the classification of plant biota and then animal biota.[2] By the twentieth century, however, researchers were mostly employed in universities and mostly specialized within a single discipline.

Universities, long in existence, were slow to become centers of research. It is true that the formal institutions of universities had existed from the early days of the Common Era in China, Persia, and in the Buddhist tradition, and that universities opened subsequently in the Islamic tradition and thereafter among Christian societies. But the early universities centered primarily on theology and the training of religious officials, and only gradually added the study of law and then medicine. For these reasons, early universities did not become centers of scientific or humanistic study.[3] By the eighteenth century, the institutions and structural conditions in which scholars could work included individual work at home, teaching in universities, seeking membership in academies, and maintaining informal networks of correspondence. Scholarship had long been mostly informal. In sixteenth-century Italy, however, a movement arose to create academies, supported by aristocrats, which provided support and recognition for scholarship. Such institutions moved north with time, resulting in the creation of royal academies in

---

[2] Carl Linnaeus (1707–1778) worked in Sweden but maintained worldwide contacts. See Kenneth Nyberg, "Linnaeus's Apostles and the Globalization of Knowledge, 1729–1756," in Patrick Manning and Daniel Rood, eds., *Global Scientific Practice in an Age of Revolutions* (Pittsburgh: University of Pittsburgh Press, 2016), 73–89.

[3] Among the earliest of secular universities was Leiden University, founded in 1575.

France (1635), England (1660), and elsewhere.[4] By the eighteenth century, the growing contacts by correspondence among scholars in Europe had led to a network that began to be called the "Republic of Letters."[5]

The rapidly changing social conditions of eighteenth-century Europe brought expansion in scholarly knowledge but also its disruption and even destruction. British and Dutch merchants established great influence in the commerce of every ocean basin; Britain and France fought recurring wars for naval dominance. The French Revolution, followed by the European conquests of Napoleon, shattered old regimes in much of Europe. After the wars, Britain gained worldwide naval and commercial dominance. Nevertheless, France, Germany, the United States, and other national communities created platforms for rapid growth and change. The expansion of knowledge in these competing nations was a key element of overall global transformation. National and imperial competition spread worldwide; a capitalist economic order formed in northwestern Europe in the seventeenth century and, by the late eighteenth century, had extended tentacles to many areas of the world.

While the achievements of scholars in the eighteenth and early nineteenth century reflected an acceleration in the production and exchange of knowledge, there was reason to search for an improved institutional structure. As the knowledge required to keep up with current developments continued to grow—be it in botany, chemistry, or mechanics—there was steady encouragement for scholars to become more specialized. Contrary tendencies continued to be important, however: the five-year field trip of Alexander von Humboldt to the Americas, 1799–1804, led him to publish an eventual 21 multidisciplinary volumes—working from his office in Berlin.[6] In the paragraphs to follow, I trace these contending tendencies in scholarship of the early nineteenth century, to show at once the acceleration of knowledge, the varying personal and institutional situations of the researchers, and the steady pressure to formalize the new research in disciplinary frameworks. In the course of this half-century of active research, it became clear that formation and expansion of universities provided the best hope for giving support to the emerging disciplines.

Studies of chemistry had advanced impressively in the late eighteenth century with the work of Lavoisier in France: he isolated oxygen and developed

---

[4] D. S. Chambers and F. Quiviger, eds., *Italian Academies of the Sixteenth Century* (London: The Warburg Institute, University of London, 1995). Leading academies and their dates of founding included the Académie française 1635, Royal Society (England) 1660, the Spanish Real Academia Española (1713), Russian Academy of Sciences 1724, Royal Dutch academy of sciences 1808, U.S. National Academy of Sciences 1863, and International Association of Academies 1899.

[5] See the online representation in the Stanford Republic of Letters: republicofletters.stanford.edu/publications. For intellectual exchanges across Eurasia in the second millennium CE, see Manning and Own, *Knowledge in Translation*.

[6] L. Kellner, *Alexander von Humboldt* (London: Oxford University Press, 1963). On the scientific implications of Alexander Humboldt's work, see Jessica Ratcliff, "The Great Data Divergence: Global History of Science within Global Economic History," in Manning and Rood, *Global Scientific Practice*, 237–54.

a system for naming chemical substances; the substance of uranium, discovered in 1789, was labeled within that system. As new chemical elements were identified, English chemist John Dalton offered an 1806 atomic theory: he argued that each chemical element was formed of distinct and irreducible atoms, which were able to connect and form compounds. Soon thereafter in Turin, Amedeo Avogadro extended the theory to allow for the existence of molecules in gases. Yet the theory was disputed for a century.[7]

The discipline of geology formed and took leadership in scientific study during the early nineteenth century. Charles Lyell's 1830 textbook argued convincingly that geologic processes were mostly gradual and iterative, rather than cataclysmic as was previously argued.[8] Careful observers in Europe and beyond identified the various strata of rock laid down over time and, at a series of meetings in the 1840s, they agreed on the names and the order of the strata from the earliest to the most recent. Attention to the fossils in the various strata was essential to ordering the strata, so that the fields of biology and geology were thus joined tightly. This work, since revised only modestly, established the principles of determining stratigraphy and demonstrated conclusively that the Earth had existed for millions of years. Thus, geologists established firmly the principles of ordinal time—the order if not the exact chronology of succeeding events and principles. These principles in ordinal time, sometimes including named periods, were applied in linguistics, biology, astronomy, and other fields. In addition, new thinking in geology had an impact on economic thinking: as geologists turned from emphasizing crises to gradual processes and thought more about great quantities of coal underground rather than scarce woodlands on the surface, economists lost interest in scarcities as well, turning natural resources into rarely adjusted parameters rather than active variables.[9]

The field of physics developed at an accelerating pace during the nineteenth century, opening new areas of analysis after the earlier focus of Isaac Newton on mechanics, gravity, and optics. Michael Faraday's experiments in England led to invention of the first electric motor in 1821. By 1844, this new knowledge had been applied with the invention of the telegraph by Samuel Morse in the United States: telegraphs spread rapidly across land and

---

[7] Amedeo Avogadro, in Turin, revised Dalton's theory in 1811 by making the distinction between atoms and molecules. Avogadro's Law states that the relationship between the masses of the same volume of all gases (at the same temperature and pressure) corresponds to the relationship between their respective molecular weights. Hence, the relative molecular mass of a gas can be calculated from the mass of a sample of known volume. Cyril N. Hinshelwood and Linus Pauling, "Amedeo Avogadro," *Science* 124 (1956): 708–13.

[8] Charles Lyell, *Principles of Geology*, 3 vols. (London: J. Murray, 1830–1833). The biologist Georges Cuvier had earlier argued that geological change was principally characterized by catastrophic events.

[9] Jean-Baptiste Fressoz and Christophe Bonneuil, "Growth Unlimited: The Idea of Infinite Growth from Fossil Capitalism to Green Capitalism," in Matthias Schmelzer and Iris Borowy, eds., *History of the Future of Economic Growth* (London: Routledge, 2018), 52–68.

sea in the era of railroads and steamships. By 1850, researchers working with gases—focusing on the generation and transfer of heat—had established two principles. That is, heat cannot move automatically from a cool body to a warmer body; and heat tends inherently to change to a more random distribution. These principles soon became generalized and identified as the First and Second Laws of Thermodynamics.[10]

Developments in biological studies, focusing on classification, proceeded slowly toward the understanding of larger patterns. Jean-Baptiste Lamarck (1744–1829), who proposed that evolutionary change took place because the physical changes in an adult organism could be passed on to the next generation, gained both praise and critique for his hypothesis. The young Charles Darwin, who studied at Edinburgh and Cambridge Universities, completed his BA at Christ's College, Cambridge in 1831. Within a few months, he had begun a five-year tour as a self-financed researcher on the *Beagle*, a vessel that circumnavigated the world, collecting samples and notes. Darwin carried with him a copy of Lyell's geology text. The world, as seen from England in his lifetime, was a place of great turmoil and change, including the age of British hegemony abroad and of industrialization at home. Within his own area of focus, the world of knowledge, it was also a time of exciting acceleration, debate, and advance. On his return, Darwin lived a private life of study and correspondence. While he became a correspondent of the Geological Society, he lived the life of a gentleman scholar and never held a university appointment. In the wake of Darwin's books and the debates on them, however, the field of biology and its subfields rapidly became fully represented in the expanding community of universities.

Sociology, in a time when it was tied closely to the field of philosophy, gained its first major statement in the three-volume *Cours de Philosophie Positive*, published by Auguste Comte in 1830. Nineteenth-century intellectual life gave initial primacy to "positivism," to use the term adopted by Comte, which emphasized breaking problems into small units and analyzing each unit in isolation.

The Emergence of Universities

In the long transformation of universities, an important step was the founding, in 1809–1810, of the Friedrich Wilhelm University of Berlin, under the leadership of Wilhelm von Humboldt.[11] Following the resounding military defeat of Prussia by Napoleon at Jena, Prussian King Friedrich Wilhelm

---

[10] Rudolf Clausius, in 1850, published an article with effective statements of the first and second laws of thermodynamics; in 1865 he published a mathematical definition of entropy.

[11] The university closed at the end of the Nazi regime in 1945; in 1949 it reopened, under the German Democratic Republic, and now continues as the Humboldt University of Berlin, in honor of the brothers Wilhelm and Alexander von Humboldt.

III supported a program of comprehensive reform of the kingdom. He appointed Humboldt, a distinguished linguist, to lead in creation of a systematic structure of Prussian education from primary schools to Gymnasiums (secondary schools) with student exams, feeding into a university that was to focus on research and to linking teaching closely to new research results. In this atmosphere, the emphasis was to be on academic freedom to pursue knowledge, rather than curricular domination by state or religious authorities. Humboldt moved rapidly to set up the full schooling system, focusing especially on the university. Yet he left his position shortly before the university formally opened in 1810, in protest over insufficient endowment of the university.[12] Universities gained some administrative independence but not financial independence.

The Friedrich Wilhelm University of Berlin opened with a traditional list of faculties: in law, medicine, theology, and philosophy. With time, however, the range of appointments expanded to most areas of active research, and the university appointed leading researchers to professorial positions. The philosopher Johann Gottlieb Fichte was appointed professor of philosophy in 1810 and was succeeded by G. W. F. Hegel from 1818 to his death in 1831. The historian Leopold von Ranke, noted for his research in the diplomatic archives of Venice, was appointed professor of history in 1825 and held his position for 50 years. Johannes Peter Müller filled a chair in anatomy and physiology in 1833; leading figures in mathematics, physics, and chemistry were to follow.[13] The strength of the university grew steadily; other universities in Germany and abroad pursued this model.

The idea caught on. Universities formed and transformed to fit what became known as the German model—for instance, the University of Helsinki was founded in 1828 in Russia's Grand Duchy of Finland. Universities began with organization into faculties, as in Berlin, but departments gradually developed within the faculties and became the homes of academic disciplines. Regional and national competition among universities mirrored growing economic competition, with formation of scientific societies and research publications. In the same process, universities became secular in at least two senses. That is, universities increasingly admitted students of religions outside the state religion (and some women and commoners), and

---

[12] Alexander von Humboldt went on to devote his later years to cataloging the data he had collected. Wilhelm von Humboldt held his position as founder of the university for only a year. Both of the Humboldt brothers completed their university studies in Göttingen. Johan Östlung, trans. Lena Olsson, *Humboldt and the Modern German University* (Lund: Lund University Press, 2016).

[13] The competing model of higher education was that of the Grandes écoles of France, which had shown their strength during the Napoleonic wars. These institutions were directed far more tightly by the state, with the ministry rather than the professor setting the syllabus. See Willis Rudy, *The Universities of Europe, 1100–1914: A History* (Rutherford, NJ: Fairleigh Dickinson University Press, 1984), 100–16.

the disciplines were increasingly able to be independent of religious orthodoxy. The principle of academic freedom—with curriculum set by professors rather than by state or religious authority—gained increasing recognition.[14]

THE TRANSITION TO DISCIPLINARY ORGANIZATION OF KNOWLEDGE

By the mid-nineteenth century, with discoveries in chemistry, physics, and geology—and also with debates in the emerging social sciences—disciplines were coalescing within their new boundaries, while the general issue of *scale* in the universe began to come to the fore within each discipline. These were efforts to see beyond the world of our immediate perception to additional forces and patterns that are not readily visible, because their scale was too small or too large to see.

Disciplines became stronger, more clearly secular, and more specific, as indicated by the academic journals they published and the courses of study they offered to those entering each field. Many of the directions of disciplinary knowledge have produced devices and techniques that have now become methods for historical analysis. Analysis in each discipline alternated among scales—they debated priorities among the scales, developed theories for each scale, and sought to create theories linking the scales (for instance, the differing scales of mechanics, thermodynamics, and electricity and magnetism in physics). They built patriotic feelings in defense of each discipline, resulting in academic walls that raised the costs of entry into each field. Nevertheless, discoveries in one discipline not uncommonly became valuable in other disciplines. The specialization of disciplines meant that parallel concepts needed to be reinvented in multiple disciplines—thus, the notion of equilibrium spread across disciplines. As an example of the expansion in universities, in 1862 the United States began awarding federally controlled land to states for them to endow land-grant colleges, thus providing a relatively sound fiscal basis for public higher education.[15] Most of these colleges became comprehensive state universities. The first American university constructed on the German model of a research university was the Johns Hopkins University, founded in 1876. In a parallel move, the University of Tokyo formed in 1877.

My narrative of disciplines and disciplinary knowledge, from this point on, moves through five periods from 1850 to 2020. The discussion moves from discipline to discipline, giving primacy to biology and the social sciences, but maintaining attention to the physical sciences and to broader social outlooks.

---

[14] Rudy, *Universities of Europe*, 113–16. Universities, as they expanded, gradually recognized academic roles beyond those of the single professor for each department; existing and newly founded Catholic universities adapted to the transformation, adding disciplinary departments but maintaining the centrality of religious study.

[15] Also in the United States, the National Academy of Sciences, created in 1863, reflected an effort to match the academies of Europe.

In addition, the concluding section of each chapter explores the shifting discourse on human evolution. This discourse, linking biological, cultural, and social evolution, linked and opposed the differing versions of social evolution and has inevitably included competing visions of progress. The combination of disciplinary narratives with the developing discourse on nature of human evolution should provide readers with critical background on the numerous historical methods presented, in different ways, in Part I and Part II of the book.

CHAPTER 9

Natural Selection in an Imperial Era, 1850–1945

The century from 1850 to 1945 was the high mark in European global hegemony. It was a century of industrialization, capitalism, and empire, including the rise of new powers within that framework; universities expanded throughout this period. The chapter is organized into two periods: the first highlights the life and contributions of Charles Darwin up to his death in 1882. The period 1850–1880 was an era of industrial growth powered by coal. It was a time of the formation of national governments in Europe and beyond, initial expansion of empires, and social conflict in Europe and abroad. New and reorganized powers arose to take their place in the world order, especially the United States, Germany, Italy, Russia, and Japan. Ideas of civilizational pride and hierarchy accompanied this Euro-American expansion.

Academic knowledge reinforced both nationalist and cosmopolitan outlooks. Great nations and small competed vigorously in the expanding capitalist economy. Institutions of national governments were strengthened, including systems of primary and secondary schools, and especially universities. Beyond the North Atlantic, Japan, China, and India also entered the world of disciplinary scholarship. The growing prestige of disciplinary knowledge led nations and even universities to control it tightly. Yet academic associations formed and met; their members understood that intellectual freedom of research and the free interchange of research findings were essential to the expansion of knowledge.

In the period after 1880, the scale of industry expanded further, now powered by petroleum. Imperial domination reached its maximal extent. As conquests engulfed Asia, Africa, and the Pacific, the imperial rulers imposed racially divisive systems of colonial rule, also sustaining ideologies of racial hierarchy and discrimination in the metropolitan lands. As wage labor forces grew in both metropoles and colonies, trade union organizations arose, many of them enunciating socialist and anarchist principles. Primary and secondary schools expanded in the metropoles if not in the colonies. Universities expanded, though women were few among those admitted. The growing number of congresses of disciplinary organizations facilitated the formation in 1899 of the International Association of Academies. In recognition of fundamental advances in the natural sciences, the initial Nobel Prizes were awarded in 1901.

THE ERA OF DARWIN, 1850–1880

Beyond the world of immediate human perception, each discipline encountered forces and patterns that were not readily visible, because their scale was too small or too large to see. In geology, economics, and other fields, the scope of analysis gradually expanded both to larger and to smaller frames, so that choices arose in each discipline as to which scale should hold primacy within the discipline, or whether its theory and research might encompass many scales.

The issue of analytical scale made itself felt in a particular fashion within the social sciences: whether the unit of analysis was to be individual humans or groups of humans. For the post-1850 era, group thinking was commonly applied at the macro-level to support racial and civilizational hierarchy. For large-scale and civilizational analysis, group categorizations were applied that distinguished populations thought to be advanced from those thought to be primitive. For smaller-scale analysis within populations thought to be advanced, individual-level analysis was employed. Thus, Arthur de Gobineau emphasized differences among races in his analysis of human inequality, while John Stuart Mill focused on individual analysis in his exploration of English economic life. At another level, Darwin analyzed animal life at the individual level, while Lewis Henry Morgan's anthropological investigations focused on hierarchies of human groups.[1] In public discourse, there was a strong tendency to lump people into categories and groups, in discussing either the present or the past. Such categorization—by race, ethnicity, gender, nation, empire, colonies, religion, and civilization—also extended to distinctions within ethnicities by urban and rural, rich and poor. Yet the meanings of categories changed with time, as did the allocation of people among categories.

A more precise version of the issue of scale was the rise in several disciplines of population studies, which explained each unit in terms of its constituent subunits. In a parallel for studies of human society, the philosophy of individualism highlighted the analysis and the agency of each individual. While population analysis grew in human studies, analyses of interactions grew only slowly. And while much social analysis treated humans in groups, the analyses of how groups functioned or related to individuals remained

---

[1] Gobineau, *Inequality of Human Races*; John Stuart Mill, *On Liberty* (London: John Parker and Son, 1859); Darwin, *Origin of Species*; Lewis Henry Morgan, *Ancient Society; Or, Researches in the Lines of Human Progress from Savagery, Through Barbarism to Civilization* (London: Macmillan, 1877).

scattered and inconsistent. In sum, I narrate a slow convergence in the paths of natural science and social science to the point where the two camps can now hope to collaborate in a comprehensive study of human evolution.

Physical sciences. The discipline of physics included leading researchers in several European countries, devoting their labs especially to work on gases and thermodynamics. Physicist William Thompson, in Glasgow, was able to calculate in 1848 that the absolute zero temperature, at which there was neither heat nor energy, was −273°C. (Thompson became Lord Kelvin in 1892.) James Clerk Maxwell, a Scottish theoretical physicist, developed a coherent theory linking electricity and magnetism, in work adding a parallel dimension to Newton's theory of motion in larger masses. Maxwell's 1861 equations analyzed the motion of the positive and negative charges of electricity and magnetism, showing the workings of electromagnetic force as both particles and waves. His work theorized the speed of light in a vacuum as invariant in any frame of reference.[2]

Practical advances depending on physics included electric engines from the 1830s, internal combustion engines from the 1870s, and turbines in the late nineteenth century, mixing steam, electricity, and internal combustion into compound forms of power. Demand for petroleum grew rapidly with the rise of internal combustion engines. In another major summary, Russian chemist Dmitri Mendeleev pulled together the work of many independent chemists who had discovered and classified chemical elements: in 1869, Mendeleev published a "periodic table of the elements," showing the known 69 elements in terms of their atomic number, atomic weights, their characteristics, and the groups into which they could be classified.[3]

Biological science. The field of biology expanded at both macro- and micro-levels in the early nineteenth century. At the micro-level, cells and their elements came to be described in work with microscopes from the 1830s. The first observation of mitosis—the division of cells in plants and animals—took place in the 1870s. But the advance in overall conceptualization came in the 1850s, based on wide-ranging field studies and long reflection by two otherwise peripheral researchers. Charles Darwin turned 41 at mid-century in 1850; he published his great book at age 50 and lived a life of prominence until his death in 1882. At the level of the organism, Darwin and Alfred Russel Wallace had independently developed the theory of natural selection by the 1850s.[4] They jointly published an article before Darwin published his 1859 book, *On the Origin of Species by Means of Natural Selection*. Through numerous examples, the book conveyed his vision of variation within species,

---

[2] James Clerk Maxwell, "A Dynamical Theory of the Electromagnetic Field," *Philosophical Transactions of the Royal Society of London* 155 (1865): 459–512.

[3] Numerous versions of this periodic table were circulated for years until a standard version emerged.

[4] Peter J. Bowler, *Evolution: The History of an Idea*, 3rd ed. (Berkeley: University of California Press, 2003).

competition in a "struggle for existence," and inheritance as the elements of natural selection.[5] In one dimension of the response to Darwin's publication, German zoologist Ernst Haeckel asserted, in 1866, that ontogeny (the embryology and life-course development of the individual) generally followed or replicated phylogeny (the developmental history of the species). Haeckel advocated his theory vigorously as the basis for embryology, but it remained even more controversial than Darwin's theory of phylogeny, the development of species.[6]

Darwin's succeeding book, *The Descent of Man* (1871), extended his earlier argument that humanity was one more in the long line of species that had emerged through natural selection: he discussed the "mental powers," the "moral faculties," and the "races of man."[7] He reaffirmed the importance of individual-level natural selection in the rise of humanity, arguing that human "intellectual faculties have been gradually perfected through natural selection," and he argued for the importance of imitation as a mechanism for human change.

Darwin went on to venture into the evolution of social groups. In his 1871 book, he described the ways in which adventurous and warlike humans, in building preparations for battle, relied on praise and blame to gather support from other members of their tribe.[8] In taking this step, he argued for a mechanism of evolution relying on interactions among members of a group rather than on individual change. He sought to use natural selection as the mechanism for change at this group level, arguing that the tribes displaying the greatest bravery, as built up through praise and blame, would survive preferentially. But by shifting from the logic of individual change to one of group change, he added to the complexity of his argument.[9] Darwin later backed away from group retention of social innovations, while his co-theorist Alfred Russel Wallace supported the notion. This issue of group evolution has since remained in debate.[10] Darwin addressed yet another biological issue that also continues to be unresolved: that of emotions. In his third book, *The Expression of Emotions in Man and Animals*, Darwin treated emotions as inheritances from ancient animal ancestors rather than analyzing them

---

[5] For Darwin's descriptions of key elements in his initial volume, see: on variation, 83, 102, 202; competition, 130, 116; on struggle for existence 91; on inheritance 75, 76. Darwin, *Origin of Species* (1859).

[6] Haeckel's theory of the consistency of ontogeny and phylogeny lost all support when the genetic basis of evolutionary change was confirmed.

[7] Half of the book addresses Darwin's argument for the mechanism of sexual selection in animals—the argument that consistent patterns of sexual preference in one sex lead to development of corresponding characteristics in the other sex throughout the animal kingdom, including in humans. Charles Darwin, *The Descent of Man and Selection in Relation to Sex*, 2nd ed. (New York: A. L. Burt, 1874).

[8] Darwin, *Descent of Man*, 163–64.

[9] Michael Ruse, "Charles Darwin and Group Selection," *Annals of Science* 37 (1980): 615–30.

[10] On group evolution, see also Chapters 5 and 6.

through natural selection.[11] There was little biological debate on emotions until the twentieth century.

Social sciences. Herbert Spencer, the self-educated English polymath, drew heavily on the individualistic and free-market philosophy of Jeremy Bentham and on the example of Comte. He sought to apply his ideas at the widest possible level and, in an 1857 article on "Progress," articulated the notion of a world in change at every level of its existence. For Spencer, the notion of "evolution," meaning an inherent tendency to change, was to be found at every level of nature and society. Quite a different notion of social change was enunciated by Arthur de Gobineau, a lawyer of French birth, whose *Essai sur l'inégalité des races humaines* divided the human species into several distinct races that could be set in a hierarchy from the most able to the least well endowed. Of the races he defined, he labeled the Aryan race as the source of all civilizations in the Eastern Hemisphere and even in the Western Hemisphere.[12] As global communications expanded during the nineteenth century, authors and readers became interested in histories of global interconnection.

The field of anthropology expanded in the mid-nineteenth century, especially as researchers in imperial centers expanded their thinking about the character of peoples under imperial rule. Edward Tylor, trained as a lawyer, read widely on peoples who were under English rule, with attention especially to those in relatively small-scale societies.[13] Lewis Henry Morgan's experience was different, in that as a young man in upstate New York he interacted with people of the Iroquois nation and became attracted to their style of life. While a lawyer by profession, he generalized his initial interest in the Iroquois to a worldwide study of *Ancient Society*, in which he expanded a previously labeled schema of stages in human development: savagery, barbarism, and civilization.[14] This stage theory became the principal form of evolutionary thinking among social scientists. The unit of analysis was the "society," whether a small ethnicity or an empire; the mode of analysis was phenotypical, in that it focused on visible, overall characteristics of the society rather than pursuing a systematic search for parallel elements of societies.

---

[11] Charles Darwin, *The Expression of Emotions in Man and Animals* (London: John Murray, 1872). Darwin briefly discussed emotions and feelings in his 1871 book: Darwin, *Descent of Man*, 158–62.

[12] Spencer, "Progress"; Gobineau, *Inequality of Human Races*. On Gobineau, see Jonathan Marks, *Human Biodiversity: Genes, Race, and History* (New York: Aldine De Gruyter, 1995), 64–66, 68–70, 73–74.

[13] Tylor, *Primitive Culture*; see also Tylor, *Researches into the Early History of Mankind and the Development Of Civilization*, 2nd ed. (London: J. Murray, 1870); Tylor, *Anthropology: An Introduction to the Study of Man and Civilization* (London: Macmillan, 1881).

[14] The schema was earlier articulated by Adam Ferguson in *Essay on the History of Civil Society* (Edinburgh, 1768). Morgan, *Ancient Society*. See also Frederick Engels, "The Origin of the Family, Private Property, and the State," in *Karl Marx and Frederick Engels: Selected Works* (New York: International Publishers, 1969), 468–593.

Thus, "evolutionism" in anthropology and other social sciences preserved a clear difference from "evolutionism" in biology.[15]

For analyses of human change at large social scales, the long discussion included models of progress through degrees of centralization, but without mechanisms. The main exception was Marx's model of social change in recent centuries. Karl Marx, trained in philosophy, developed the contrasting philosophy of "dialectical materialism" with an emphasis on interactions; he applied it in study of the human order. Working with Friedrich Engels, he entered political activity with the *Communist Manifesto* in 1848, emphasizing social activism from the standpoint of wage workers.[16] He argued that conflicts among economically based social classes brought a transformation from feudalism to capitalism, in which wage workers sought to create a further shift to socialism.[17] Léon Walras, another original economic thinker, focused on a cross-sectional view of economic processes, tracing the "general equilibrium" that could clear all markets. Alfred Marshall, working at a smaller scale, introduced the theory of the firm beginning in 1890.[18] The range of mid-century frameworks proposed by writers in the social sciences was parallel to the range of biological theories in the days before Darwin. But the range of social science data was temporally narrow, so that it rarely focused on specific processes of change.

The publication of universal histories, multi-volume works seeking to assemble the whole human past into edifying stories, became popular in the mid-eighteenth century and continued on a smaller scale into the nineteenth century. Nikolai Gogol, in Russia, penned an elegant and concise survey of world history in 1835. "Peter Parley's Universal history," by Samuel Goodrich, appeared in 1859.[19] Yet the rise of research universities in an era of national pride brought a divergence in historical writing. The outstanding leader in historical studies was Leopold von Ranke, who began to lead seminars at the Friedrich Wilhelm University of Berlin in 1825, in which his students were trained to read in the diplomatic archives of states, especially

---

[15] Darwinian analysis in biology also focused on the phenotype of the organism, but its analysis focused on hypothesizing a specific mechanism within the organism. This hypothesis was rewarded by 1910 with the conception of the gene and in 1944 with Oswald Avery's confirmation of nucleic acids as the genetic material.

[16] The adoption of Morganesque stages into Marxian thinking came later, beginning with Engels' 1884 "Origin of the Family, Private Property, and the State."

[17] Marx (1818–1883). Vol. 1 (1867). The remaining two volumes were edited by Engels from Marx's notes and published in 1885 and 1894.

[18] Léon Walras, *Études d'économic politique appliquée Théorie de la production de la richesse sociale* (Lausanne: Corbaz, 1874); Alfred Marshall, *Principles of Economics* (London: Macmillan, 1890).

[19] For a remarkably insightful view within universal history, see Nikolai Gogol, trans. Alexander Tulloch, *Arabesques* (Ann Arbor: Ardis, 1982 [1835]). For a widely-read survey, see *Peter Parley's Universal History on the Basis of Geography*, by Samuel G. (Samuel Griswold) Goodrich, 1793–1860. Publication date 1859.

European, to write up political and diplomatic histories. As national governments and national identity grew stronger in many parts of the world, historians focused increasingly on interpreting the nation and in following the guidelines of Ranke.[20] Universal history persisted, however, if outside the academy. Winwood Reade's *The Martyrdom of Man*, first published in 1872, went through twenty-five impressions and was still in print in 1927.[21] In Meiji Japan, several universal histories by American authors were translated into Japanese beginning in the 1870s.

Understandings of evolution by 1880. Four versions of thought about evolution—at once biological and human—had gained space on the stage for public discussion by 1880. The best established of these was the discourse on civilization, which assumed leadership and progress for those well placed in powerful societies. This approach ranked civilizations more through essentialism—the arbitrary assertion of inherent qualities in a subject—than through the identification of a logically displayed process. Spencer's attempt to generalize evolution—at levels both civilizational and universal—gained wide attention and maintained its popularity for two or three decades. The general expansion of inquiry, paired with expansion of European conquests, was sure to bring this topic into discussion.

The third strand of thought, an academic discourse on biological evolution, was new but had been presented with such strength that it rapidly gained remarkable prominence. Darwin's primary argument was that all animals had evolved in a common process, in which humans were linked to other species. Second, he asserted a mechanism of biological change: "descent with modification" through natural selection. As he argued, the underlying processes included a variety of characteristics within each species, a "struggle for existence" in which those individual organisms that were most fit (as measured by the number of their offspring) had characteristics that were inherited by their offspring. This analysis, while it included few specifics beyond the level of phenotypic description, was on its way to being accepted by consensus among biologists; it encouraged further research to seek underlying mechanisms. While Darwin did not initially use the term "evolution," by 1872 he had acceded to its wide adoption and included it as a synonym for natural selection in later editions of his books.[22]

The fourth strand of discourse, at once academic and of general interest, was the anthropological analysis of Tylor and Morgan and their contemporary colleagues. In it, tribes, states, and perhaps civilizations were seen to have

---

[20] Ranke, a historian of the state, believed that the state played a prime role in society and that individuals should strive to fulfill the "idea" of their state.

[21] Winwood Reade, *The Martyrdom of Man* (London: Jonathan Cape, 1926 [1872]).

[22] Enthusiasm for Darwin's thesis declined by the opening of the twentieth century; in the 1930s, with the clear linkage of Darwinian and Mendelian theory, Darwinian theory gained a primacy it has retained. Peter J. Bowler, *Evolution: The History of an Idea*, 3rd ed. (Berkeley: University of California Press, 2003).

undergone changes through the exercise of human consciousness. It was logical that this analysis should be compared with that of Darwin. The Darwinian focus on "fitness," measured most directly by numbers of offspring of an organism, appeared to be translatable into human terms. That is, population growth of societies could be seen as an indication of fitness and evolutionary success. Without thinking carefully about units of analysis, members of the reading public found it straightforward to take societies as analogous to species, and population growth in societies as analogous to numbers of offspring for an organism in biological evolution.

Before the turn of the century, however, the complications and the unresolved issues in each of these approaches became apparent, and critical voices arose in each arena. Nevertheless, one may argue that a relatively comprehensive discourse on human evolution arose in the era of Darwin, in which discussions of biological and social change, along with changes in the natural world, were studied and debated as never before. The apparently triumphal expansion of European-based influence throughout the world encouraged intellectual optimism, though there were war and famine enough to suggest that problems lay ahead. Looking back at this moment more than a century later, anthropologist Stephen Sanderson called it the era of "classical evolutionism."

In the Wake of Darwin, 1880–1945

Scientific work after 1880 brought important new understanding of the scales of the physical, biological, and even social universes. Statistical thinking moved from the mathematical to the real world, showing ways to treat physical and social aggregates in terms of their elements.[23] The scale of time came to be seen in detail through understanding of radioactivity; time came to be seen as flexible through the theory of relativity. Space came to be assessed more broadly through the study of geology and geography. The certainty of positivistic logic began to give way to multiple perspectives, for instance through Heisenberg's "uncertainty principle," according to which time and space cannot both be known with complete precision at the quantum level.

Despite these increases in sophistication across the disciplines in the understanding of scale, the social sciences diverged increasingly from natural sciences after 1880. As a sign of that change, the German philosopher Wilhelm Dilthey argued in an 1883 book that the human sciences should be separated formally and philosophically from the natural sciences.[24] Still, there

---

[23] Ironically, however, some of the principal advances in statistical technique, especially by Karl Pearson and Charles Spearman, were devoted to the effort to demonstrate human intellectual inequality through eugenics.

[24] Dilthey sought also to distinguish between an atomistic/mechanistic and a holistic/organismic metaphysics. Wilhelm Dilthey, trans. Ramon J. Betanzos, *Introduction to the Human Sciences* (Detroit: Wayne State University Press, 1988 [1883]), 209–18; Lars Udehn, *Methodological Individualism: Background, History, and Meaning* (London: Routledge, 2001), 357.

remained overlaps: notions of equilibrium prevailed in economics and chemistry, along with the "balance of nature" in ecology. More broadly, individualistic philosophy dominated social science thinking as it had in the previous period. This period brought expansion of empire and militarism, as well as expansion in racial discrimination, segregation, colonial hierarchy, and the hierarchical conception of civilization. The scholarly interest in social evolution at the start of the twentieth century may have been tied to enthusiasm for imperial expansion, treating European societies as "most fit." Yet collective and corporate ideologies challenged individualism, through practices of campaigns for women's rights, trade unionism, socialism, nationalism, fascism, and with heightened racial discrimination and hierarchy. Disruptions of equilibrium arose in quantum theory and in socialist and civilizational theories of social conflict. Debates continued as to the exact meaning and mechanism of biological evolution, so that eugenics gained wide support while its opponents saw it as contradictory to human values and biological processes.

Physical sciences. The field of physics was particularly brilliant in its innovations from 1880 to 1945. Practical inventions included the creation of public supplies of light and power from the 1880s and the "wireless telegraph" or radio from 1897. Ludwig Boltzmann, working in Austria especially from 1874 to 1890, deepened the work of Maxwell. He restated the interpretation of the Laws of Thermodynamics, assuming the existence of atoms and molecules as constituents of gases. His equations, known as statistical mechanics and statistical thermodynamics, related the kinetic energy of atomic particles in a gas to the temperature of the gas and stated the laws of thermodynamics in terms of atoms and molecules as well as the gas overall.[25] His work thus bridged the macroscopic and microscopic sides of physics and opened the door to similar developments in genetics, demography, and quantum mechanics. Yet the opponents of atomic theory refused for decades to accept Boltzmann's results.[26]

In 1895, the German physicist Wilhelm Röntgen produced and detected electromagnetic radiation at a high wavelength that penetrated heavy paper and other materials: he labeled this radiation as *X-rays*.[27] Henri Becquerel, a French physicist, immediately discovered that the mineral pitchblende, known to contain uranium, emitted natural rays that similarly passed through a thick layer of paper and were recorded on a photographic plate. His associates Pierre and Marie Curie found that the rays, which Marie Curie labeled as "radioactivity," came from the uranium; they also found that there was a second source of radioactivity in the pitchblende. After years of research

---

[25] Olivier Darrigol, *Atoms, Mechanics, and Probability: Ludwig Boltzmann's Statistico-Mechanical Writings—An Exegesis* (Oxford: Oxford University Press, 2018).

[26] To the end of his life in 1906, Boltzmann argued against philosophers and physicists led by Ernst Mach who refused to accept the reality of atoms and molecules.

[27] Eva Hemmungs Wirtén, *Making Marie Curie: Intellectual Property and Celebrity Culture in an Age of Information* (Chicago: The University of Chicago Press, 2015).

and processing (working in a ramshackle lab in Paris), Marie Curie purified radium and identified it as a new and radioactive element; she also identified polonium as the second new element and third radioactive element. Becquerel and the two Curies received Nobel Prizes in physics in 1903. Meanwhile, New Zealand-born physicist Ernest Rutherford, working in England, identified radiation as consisting of two types: alpha particles (which were the same as helium atoms) and beta particles (electrons); he later labeled gamma particles (photons, the particles of light, including X-rays).

Max Planck, continuing the work of Röntgen, observed in emissions of electromagnetic radiation from a black box that the emissions consisted of packets of varying energy, which he called "quanta." In discussion with Albert Einstein, he was brought to realize that these "quanta," at once particles and waves, were part of a more general phenomenon. Planck's discovery was the beginning of quantum theory, which developed steadily through the next few decades. In parallel to Planck's work, Einstein proposed that light also delivers its energy in chunks; light would then consist of little particles, or quanta, soon labeled photons, each with an energy of Planck's constant times its frequency.[28] Einstein's special theory of relativity, in 1905, linked Maxwell's theory of electricity and magnetism to Newton's theory of gravity and massive bodies. Einstein's modification of Newton's theory showed that time, rather than an invariant quantity, was a variable that depended on the perspective of the observer.[29] This result, along with the discovery of the varying energy levels of electrons circulating atomic nuclei, became part of the "quantum mechanics" as articulated by Niels Bohr of Denmark during the 1920s.[30] The work of chemists and physicists discovered various subatomic particles, confirming atomic theory yet replacing Dalton's early theory of irreducible atoms with a "planetary" model of atoms: electrons orbited about nuclei made of protons and (as was later discovered) neutrons.

Three key advances in geology clarified long-term patterns of the Earth. Alfred Wegener articulated his theory of continental drift in 1912, though it could not be confirmed until roughly 1960. Geologist Arthur Holmes drew on the observation of the physicist Rutherford that radioactivity should permit estimation of the ages of rocks. Holmes, a recent geology PhD in England, relied on analysis of lead and helium to argue, in a 1913 book, that

---

[28] Planck sought to fit equations to the observed intensity of radiation from a black body. An essential step in this work was Planck's reluctant adoption of Boltzmann's statistical mechanics. Planck's resulting law included three constants: the speed of light (c) from Maxwell, a ratio of energy to temperature (k) from the work of Boltzmann, and a ratio of energy to frequency (h), which came to be known as Planck's constant.

[29] Einstein's general relativity (1915) added acceleration to the analysis. For engaging and accessible detail on the special theory of relativity, see Walter C. Mih, *The Fascinating Life and Theory of Einstein* (Bloomington, IN: AuthorHouse, 2004), 114–21, 146–62.

[30] By 1930, quantum mechanics had stabilized: the 1935 text by Pauling and Wilson is still in print. Linus Pauling and E. Bright Wilson, Jr., *Introduction to Quantum Mechanics, with Applications to Chemistry* (New York: McGraw-Hill, 1935).

the world was at least 1.6 billion years old.[31] This work proposed chronological dates not only for the Earth but for all of its geological strata. Holmes pursued this work for the rest of his career, revising estimated ages for the Earth and its strata several times and, working with others, in the 1940s came up with the figure still used today for the origin of the Earth, 4.5 billion years. Third, Milutin Milankovitch, an astrophysicist at the University of Belgrade, studied the patterns of the Earth's orbit around the sun and developed in the 1920s a theory emphasizing three elements of that relationship: variations in the eccentricity of the Earth's orbit around the sun; changes in obliquity (the angle that Earth's axis makes with the plane of the Earth's orbit); and precession (the change in the direction of the Earth's axis of rotation). Milankovitch's theory was to predict the amount of "insolation"—the amount of solar energy—reaching each point on the Earth over time. It took almost forty years, however, until advances in geologic theory and sampling permitted his theory to be applied to measures of past climate.[32]

Biological sciences. During the late nineteenth century, medical scientists made important advances in discovering causes of disease and developed new treatments, initially by proving that bacteria, first observed by microscope in the 1680s, transmitted certain diseases. Louis Pasteur began in the 1850s to conduct experiments on the basis of the "germ theory" of disease, according to which microorganisms—rather than the vague "miasma" of the dominant theory—were the agent of infection. Robert Koch in Germany, after isolating the anthrax bacillus in 1876, articulated a set of four postulates that were very helpful for identifying specific agents of diseases. Even as the germ theory triumphed, it became clear that there were sharply different types of "germs"—bacteria, viruses, protozoa, and fungi.[33] The excitement and hope brought by work on bacteria would be succeeded by disappointment and delay elsewhere. The protozoan parasite causing malaria was correctly observed in 1880, but it took until 1897 to identify the *Anopheles* mosquito as its carrier. The first virus to be diagnosed was the infectious agent for the tobacco mosaic disease, in 1892. Yet viruses could not be observed directly for decades, until the invention of electron microscopes. The field of pharmacology began to develop, for instance as the Bayer firm began marketing aspirin as a pain medication from 1899.

---

[31] Arthur Holmes, *The Age of the Earth* (London: Harper Brothers, 1913). The initial estimate of 1.6 billion years was revised until the 1940s, with an estimate of 4.5 billion years that has remained unchanged since.

[32] Milankovitch cycles were initially traced back some 450,000 years, but the analysis has subsequently been traced back as far as the Triassic period, some 200 million years ago. Milutin Milankovitch, "Mathematische Klimalehre und Astronomische Theorie der Klimaschwankungen," in *Handbuch der Klimatologie*, 5 vols., Band 1, Allgemeine Klimalehre. Teil A. Mathematische Klimalehre und astronomische Theorie der Klimaschwankungen (Berlin: Verlag von Gebrüder Borntraeger, 1930).

[33] F. E. G. Cox, "History of Human Parasitology," *Clinical Microbiology Reviews* (2002): 595–612.

Following Darwin's work and up to 1940, other researchers proposed large and small revisions to his vision of natural selection. Theodor Eimer, in Germany, advanced in 1898 the hypothesis that organisms undergo evolutionary change in a specific direction because of some internal mechanism: he appropriated the term *orthogenesis* to represent it.[34] Meanwhile, a group of English researchers—Francis Galton, Charles Spearman, and Karl Pearson—had developed the idea of "eugenics."[35] They sought to show that human characteristics, especially intelligence, were inherited in a relatively straightforward system, so that it would be possible to breed humans and improve the intelligence of the race. Statistical measures of Spearman and Pearson, including twin studies, became important in the rise of social statistics. Nevertheless, eugenics rapidly raised suspicion that it was a reflection of social prejudice, as it seemed to call for elimination of most people in society in order to expand the offspring of those determined to be of highest quality. In the meantime, eugenics grew as a field and it came to be reinforced by intelligence testing. Galton attempted intelligence tests from 1882 without success; William Stern of Germany developed the IQ test drawing on the work of Alfred Binet in France; and a US test was applied to soldiers during World War I. But the developers tended to assume a straightforward relationship between test results and inherited intelligence, a result that later met fundamental challenge.[36]

In the new century, work on genetic inheritance became more detailed. Gregor Mendel's work of the 1860s had established the basic relations of dominant and recessive genes in sexual reproduction through his work with the colors of sweet-pea flowers. His work finally gained attention in 1900 when scholars returning to the details of inheritance encountered his publications.[37] Wilhelm Johannsen, a Danish biologist who had coined the term "gene" to refer to the unit of genetic evolution, developed in 1911 a terminology distinguishing the "genotype"—the genetic constituents of an organism—from the "phenotype" of the same organism—the outward characteristics and behavior of the organism.[38] Research came closer to identifying the nature of genes and their role in inheritance. Leaders were Thomas H. Morgan and

---

[34] "Ortho" is from the Greek for right, straight, or correct.

[35] The term "eugenics" begins with the Greek prefix "eu" for "good" or "well"—hence "good genes." Richard Lynn, *Eugenics: A Reassessment* (Westport, CT: Praeger, 2001); Aaron Gillette, *Eugenics and the Nature-Nurture Debate in the Twentieth Century* (New York: Palgrave MacMillan, 2007).

[36] Ken Richardson, *Genes, Brains, and Human Potential: The Science and Ideology of Intelligence* (New York: Columbia University Press, 2019).

[37] Botanists Hugo de Vries, Carl Correns, and Erich von Tschermak independently found Mendel's publications in 1900.

[38] Wilhelm Johannsen, "The Genotype Conception of Heredity," *The American Naturalist* 45 (1911): 129–59. https://doi.org/10.1086/279202. JSTOR 2455747. For an early and thoughtful critique of Mendelian genetics, see George H. Shull, "Mendelian or Non-mendelian?" *Science* 54 (1921): 213–16. Johannsen coined the term "gene" from.

his group in the United States. They conducted studies beginning 1910 on fruit flies, *drosophila melanogaster*, that displayed the chromosomes within cell nuclei that contained genetic material. Meanwhile, the theoretical analysis of genetics advanced rapidly with the work of Ronald Fisher who, in England from 1918 to 1930, led in developing the field of population genetics.[39]

Research strategy in life sciences began to take an approach that was then being labeled as "reductionism." The Rockefeller Foundation, active in funding biological research, decided to encourage physicists to work on questions of biological inheritance, arguing that physicists' focus on essential principles would clear a path through the complexity of biology.[40] In fact, their insight was rewarded, notably with the work of Max Delbrück, who moved from studying physics with Niels Bohr into four decades of work in microbiology. This "reductionist" idea that biological phenomena could be explained in terms of physical and chemical principles was to reappear and be debated in different forms thereafter.[41] In a synthesis of the advances in genetics, the Russian-born Theodosius Dobzhansky, who had come to work with Morgan in the United States, published a 1937 book that established what became known as a "neo-Darwinian synthesis." This approach linked Darwinian natural selection to Mendelian genetics as expressed in population genetics. Dobzhansky argued firmly for the great diversity of the genome, in contrast to the supporters of eugenics who assumed a simpler and less diverse genome. Later developments were to confirm the hypothesis of Dobzhansky.[42]

Social Sciences. The rise of professional anthropology, at the start of the twentieth century, led in different directions for various national schools. In the United States, Franz Boas, a German immigrant with an initial degree in physics but with years of study and publication in anthropology, became professor of anthropology at Columbia University in 1899, where he published actively and directed many of the leading American anthropologists for the

---

[39] Garland E. Allen, *Thomas Hunt Morgan: The Man and His Science* (Princeton: Princeton University Press, 1978); Ronald A. Fisher, *The Genetical Theory of Natural Selection* (Oxford: Clarendon Press, 1930). The related work of Sewall Wright in the United States included the concept of genetic drift and the practice of path analysis, which anticipated the importance of ontological development as well as genetic processes. See Judea Pearl and Dana Mackenzie, *The Book of Why: The New Science of Cause and Effect* (New York: Basic Books, 2018).

[40] At the same time, the Rockefeller Foundation made a substantial award to the Kaiser Wilhelm Institute of Anthropology, Human Heredity, and Eugenics in Berlin, soon after the institute was founded in 1927.

[41] Michel Morange shows that efforts to reduce biology to physics and mechanics began in the seventeenth century, but argues that, "in the middle of the twentieth century, [reductionism] swept biology to the feet of the structural chemists." Michel Morange, trans. Matthew Cobb, *A History of Molecular Biology* (Cambridge, MA: Harvard University Press, 1998), 5, 243–246.

[42] With time, the term "modern synthesis," proposed by Julian Huxley, came to be used as well as "neo-Darwinian synthesis" Theodosius Dobzhansky, *Genetics and the Origin of Species*, introduced by Stephen Jay Gould (New York: Columbia University Press 1982 [1937]); Julian Huxley, *Evolution: The Modern Synthesis* (New York: Harper & Bros., 1942).

generation to come.[43] By the time that Boas began his research, the earlier enthusiasm for Herbert Spencer—support for his vision of progress and evolution—had declined, as the complications of imperial competition and racial hierarchy were becoming clearer. Boas argued explicitly against categorization and hierarchy by race. He argued equally firmly against evolutionary and hierarchical categorization of ethnic and social groups, associated with the notion of social progress. (The term "evolution," as he encountered it in its usage within anthropology, assumed a progression of societies, commonly from stage to stage, but without a mechanism for change parallel to natural selection.) Boas advanced the concept of "cultural relativism" to argue that no society could be classified as more modern or progressive than others, and he thus contested the notion that societies relying on hunting and gathering could be labeled as "our contemporary ancestors," as was sometimes done.[44] To phrase it differently, Boas rejected the notion of "orthogenic evolution," in which evolution was self-generating.[45] He advanced a vision of human nature that depended overwhelmingly on culture at the community or social level and thus on the conscious constructions of humans. Boas did not strictly reject every notion of social evolution but set the bar very high in his demand for clear processes that could sustain evolution. By the 1940s, several students of Boas, by now senior scholars, generalized their specific ethnographic research into broad interpretations of the human order: during World War II, Ruth Benedict, a prominent former student of Boas, wrote *Race: Science and Politics* as a public statement on race, especially as a critique of the racial hierarchy emphasized by the Axis powers in their war effort.[46]

[43] Charles G. Seligman and Bronislaw Malinowski (at London School of Economics) directed a parallel set of British social anthropologists of the early twentieth century. American, British, and other social anthropologists studied social groups in that they devoted great effort to kinship theory.

[44] For thorough evaluations of Boas's thinking, see Herbert S. Lewis, "Boas, Darwin, Science, and Anthropology," *Current Anthropology* 42 (2001): 381–406; and Lewis, "The Individual and Individuality in Franz Boas's Anthropology and Philosophy," in Regna Darnell, et al., eds., *The Franz Boas Papers*, Vol. 1 (Lincoln, NB: University of Nebraska Press, 2015), 19–41. For an additional evaluation, including Boas's view of human nature, see Degler, *In Search of Human Nature*, 59–104. See also Franz Boas, *The Mind of Primitive Man* (New York: Macmillan, 1938 [1911]); Boas, *Anthropology and Modern Life* (Westport, CT: Greenwood, 1984 [1928]); Marks, *Human Biodiversity*, 20–21, 71–76.

[45] Note that the meaning of "orthogenic" was somewhat different in meaning in biology and anthropology—in biology, orthogenic (or later epigenetic) change meant life course development of individuals through the action of proteins; in anthropology, orthogenic meant cultural change from generation to generation through conscious action of human organisms.

[46] Alfred Kroeber, another student of Boas, emphasized a separation of cultural from biological analysis, and at times used the term "civilization" as a synonym for "culture." Trigger argues that Boasian "historical particularism" dominated social anthropology to the 1960s but notes that studies of physical anthropology were dominated by a focus on race. Degler, *In Search of Human Nature*, 90–101; Alfred L. Kroeber, *The Nature of Culture* (Chicago: University of Chicago Press, 1952); Bruce Trigger, *Sociocultural Evolution: Calculation and Contingency* (Oxford: Blackwell, 1998); Ruth Benedict, *Race: Science and Politics*. New York: Modern Age Books, 1940.

Social scientists formed academic disciplines in this era, focusing most of their work on societies after 1500 CE: anthropologists studied small, rural societies, and other social scientists studied large-scale societies. Microeconomic theory developed through the work of British and continental scholars. The Italian scholar Vilfredo Pareto took up the study of economic equilibrium, as initiated earlier by Léon Walras in Switzerland, and published in 1896–97 a two-volume book that included identifying the conditions under which an equilibrium might be established in markets. As Pareto noted, the conditions required perfect competition and perfect information, so that equilibria might not be achieved in most cases. By 1909, Pareto had switched from economics to sociology, focusing on the circulation of elites and the question of why economic equilibria were rarely achieved.[47] These principles of economics and sociology, which drew in part on Adam Smith's notion of that expression of self-interest can lead to the greatest social good, were phrased as cross-sectional terms rather than in terms of historical sequences. Larger-scale economic analysis arose after World War I with national income analysis, input-output analysis, and John Maynard Keynes' macroeconomic theory of income and employment, which encouraged government borrowing to expand and adjust production, income, and employment.

For a time, sociologists explored the relationships among different types of social groups: Ferdinand Tönnies distinguished *Gemeinschaft und Gesellschaft* or community and society; Emile Durkheim, in France, considered similar groups.[48] Max Weber turned then to large-scale social organization and the nature of bureaucracy. In 1904, he published (in German) *The Protestant Ethic and the Spirit of Capitalism*, emphasizing religious outlook as providing the essential impetus for socioeconomic change. He pursued studies of religion and the social order with subsequent books on China, India, and Judaism. Weber, along with his brilliant student Joseph Schumpeter, wrote on large-scale issues but with a firm commitment to methodological individualism as the key to human behavior. This approach was in explicit contrast to the contemporary views of Franz Boas on cultural relativism and group

---

[47] Vilfredo Pareto, trans. Ann Schwier, *Manual of Political Economy* (New York: Augustus M. Kelley, 1971 [first published 1906, translated from French edition of 1927]), 261, 451–62, 475. See also Pareto, *Cours d'Économie Politique*, 2 vols. (Lausanne: F. Rouge, 1896–1897); Pareto, *Trattato di Sociologia Generale* (Florence: G. Barbèra, 1916) [English translation *The Mind and Society*, trans. Andrew Bongiorno and Arthur Livingston (New York: Harcourt, Brace, 1935)]. Economic equilibria had been hypothesized in 1838 by Cournot and were developed in detail by Walras in 1874.

[48] Ferdinand Tönnies, *Community and Society*, trans. and ed. Charles P. Loomis (New York: Harper and Row, 1963 [1887]); Émile Durkheim, *De la division du travail social* (Paris: Alcan, 1922 [1893]).

behavior.[49] From this base, Weber turned in his later work to a general analysis of economy and society.[50]

The field of demography, usually classified as a subfield of sociology, developed especially in government service, where the design and implementation of census tabulations expanded especially in the late nineteenth century in all countries that were highly literate. Theorists of demography, meanwhile, were scattered among several disciplines. The development of theoretical statements came in the early twentieth century, at much the same time as population genetics in biology. Alfred Lotka, an independent scholar in the United States, wrote founding texts in demography and what became systems theory, notably in the 1920s.[51]

The discipline of history formed its professional associations in the 1880s, though history remained divided into professional history (focusing on national history and aimed at academic audiences) and amateur history (in which skilled authors wrote for broad audiences on biography, politics, and universal history). Following the global conflict of the Great War, 1914–1918, the tradition of universal history written beyond the academy reappeared: erudite authors responded by producing global analyses reconsidering the human condition and the direction of human social change. Oswald Spengler produced a study of civilizational rise and fall focusing on elite groups, while H. G. Wells composed an eclectic narrative from the creation of the Earth into the future. Arnold J. Toynbee wrote a ten-volume history on the experience of successive societies of civilizational scale.[52] In a somewhat different response to wartime cataclysm, the *Annales* group of professional

---

[49] Udehn, *Methodological Individualism*, 95–107; Joseph Heath, "Methodological Individualism," *The Stanford Encyclopedia of Philosophy*, Summer 2013 Edition, Edward N. Zalta (ed.), https://plato.stanford.edu/archives/sum2013/entries/methodological-individualism/; Boas, *Mind of Primitive Man*; see also Vilfredo Pareto, *The Rise and Fall of Elites: An Application of Theoretical Sociology*, introduced by Hans L. Zetterberg (New Brunswick: Transaction Publishers, 1991).

[50] Much of Weber's scholarship was published and translated with some delay and, because he died before completing *Economy and Society*, the completion, publication, and translation of that manuscript took until the 1960s. Thus, Weber had influence on several generations of scholars. Max Weber, Guenther Roth and Claus Wittich, eds., *Economy and Society: An Outline of Interpretive Sociology*, 3 vols. (New York: Bedminster Press, 1968).

[51] Alfred J. Lotka, *Elements of Physical Biology* (Baltimore: Williams & Wilkins, 1925).

[52] This phenomenon, the appearance of global historical studies in the aftermath of large-scale war, can be observed at various times: the end of World War II, the end of the Napoleonic Wars, the end of the 30 Years War, and arguably others. Oswald Spengler's *Decline of the West* (1918) was a cultural history of great civilizations; H. G. Wells' *Outline of History* (1920) offered a narrative from the creation of the earth into the future; and Arnold J. Toynbee's *A Study of History* appeared in 12 volumes, half of them after each of the world wars—Toynbee focused on the rise and fall of "societies" or civilizations, focusing on a dynamic of challenge and response. Oswald Spengler, trans. Charles Francis Atkinson, *The Decline of the West* (New York: A. A. Knopf, 1926 [1918]); H. G. Wells, *The Outline of History* (London: George Newnes, 1920); Arnold J. Toynbee, *A Study of History*, 12 vols. (Oxford: Oxford University Press, 1933–61).

historians in France, beginning 1929, founded an academic journal and a school of historians seeking to study human society at many levels.[53]

Understandings of evolution by 1945. At the end of World War II, public discourse on growth and evolution gave much less attention to hopes for progress and the advance of civilization than had been the case in 1880. The unfortunate histories of global economic depression, the casualties and massacres of two world wars, and the conflicts of fascism, communism, and capitalism had brought doubt about civilization. German academia, which had led the world in many fields for a century, had been weakened by defeat in World War I and was almost totally dismantled by 1945. Within the ivory towers that were still standing, scholars found that caution was reinforced by factors more specific to their disciplines.

In academic studies of society in anthropology, interpretations based on the growth of hierarchy and ranking of societies had been challenged by the Boasian defense of the value of each society in its own terms. Social science research of the early twentieth century focused on learning more about the functioning of small-scale societies, though the colonial regimes within which those peoples lived continued to oppress and exploit them. For instance, British social anthropology, under Malinowski and Radcliffe-Brown, emphasized study of clan and lineage structures in societies without states.[54] In sociology, Tönnies and Durkheim wrote of human group behavior but in vague terms; Weber followed by emphasizing individual behavior as the basis of all large social structures. Among universal historians, Spengler and Toynbee each proposed a long-term advance in civilizational sophistication, though with cyclical fluctuations. Most other historians and sociologists limited their studies to change within nations and empires that had developed by the sixteenth or seventeenth century. Anthropologists considered somewhat longer periods of time and, therefore, somewhat smaller social orders. For all of these social scientists, the vision of society was overwhelmingly phenotypical; it involved little consideration of mechanisms within societies of a sort that might have been inspired by Darwinian thinking.

In biological studies, outlooks were commonly positive, but the optimism had little to do with possibilities for progress or advance in biological evolution. Instead, the discoveries in biology centered on small-scale details of the functioning of living systems: chromosomes, genes, dominant and recessive genes, and the initiation of work on the constituents of RNA and DNA. Darwin's theory was then further reaffirmed by Dobzhansky's 1937 neo-Darwinian synthesis that specified the mechanism of genetic reproduction with a focus on genes. Genes came to be seen as chemical constituents that were organized into chromosomes held in the nucleus of cells, and which

---

[53] *Annales d'histoire économique et sociale*, edited by Lucien Febvre and Marc Bloch.

[54] Adam Kuper's critical review spoke of "the invention of primitive society." Kuper, *The Invention of Primitive Society: Transformations of an Illusion* (London: Routledge, 1988).

governed heredity through processes involving cell division. Still, the precise chemical constituents of genes and genetic change remained a mystery.

Eugenics, the main hope for progress in genetics, was increasingly discredited, especially as the high degree of diversity in the human genome was documented increasingly, thus limiting the eugenic hope of selecting the best. Dobzhansky's focus on the diversity of the genome countered efforts among psychologists to identify straightforward heritability of intelligence (and other characteristics that could clearly be ranked), so that breeding appeared increasingly unlikely to lead to valuable results. Experiments with breeding and with sterilization of humans, individually cruel, were brought to a halt only after World War II.

CHAPTER 10

DNA in a Progressive Era, 1945–1980

The world's most wide-ranging and destructive war, fought by the Axis powers on a policy enforcing racial hierarchy, ended with surrender in Germany, then with another surrender after atomic blasts destroyed two Japanese port cities. Opposition to large-scale war remained widespread for decades. Decolonization took place in waves from the 1940s to the 1970s, ending the imperial era. Postwar rejection of racial categorization was influential, as seen through decolonization, although racially focused backlash arose repeatedly. In nations around the world, public investment reached peaks in the era up to 1975, expanding education, health care, and pensions. In the midst of this era, a wave of social movements protesting governmental policies shook the world, especially in 1968. Throughout, the surviving big powers, the United States and the Soviet Union, carried on a four-decade Cold War confrontation.

The postwar era was a time of dramatic advance in many arenas of academic knowledge. The scale of space came to be appreciated in much greater generality: the physical understanding of plate tectonics reaffirmed the social changes of decolonization and formation of the United Nations, as well as the academic innovation of area-studies scholarship. Multidisciplinarity began to claim some attention through systems theory and the rise of ecology. In temporal terms, however, information on evolution continued to be held in separate containers for the Pliocene, Pleistocene, and Holocene eras and for modern times. Even within these temporal limits, however, interest in social evolution arose in association with decolonization and support for group welfare in education and health. This academic interest in social welfare began to weaken in the 1980s, as ideological winds shifted toward individualism and to the criticism of social welfare projects.

Systems analysis. Norbert Wiener led in the development of cybernetics—mathematical formulations of systems that had immediate impact in electrical engineering. Swiss biochemist Ludwig von Bertalanffy had been developing a general approach to systems theory in the interwar years, although he published a book version of it in the United States only in 1969.[1] Beginning in the mid-1940s, the Smithsonian Institution gathered scholars from several disciplines for a recurring discussion on cross-disciplinary knowledge in the social sciences: William J. Baumol in economics, Talcott Parsons in sociology, and James G. Miller in biology were included.[2] In game theory, a series of competitive options including two players, John von Neumann published a 1944 book, setting forth a set of problems and solutions, to which John Nash responded with modifications in 1951. Nash's notion of an equilibrium, when applied in biology, led to identification of an "environmentally stable strategy" (ESS) that could be sustained by natural selection; later, the idea was applied in studies of cultural evolution. A third framework, field theory, arose with the 1947 articles on field theory by psychologist Kurt Lewin: this approach examined interactions between the individual and the surrounding field or environment.[3]

Even the idea of modernization had its early ties to systems analysis, through the participation of Talcott Parsons in the Smithsonian Institution's cross-disciplinary discussions. Parsons, who had previously shown great interest in the sociological work of Max Weber, led in developing "modernization theory" out of Weber's systemic approach to sociology. This theory divided societies into those following modern practices, especially at their elite levels, and those following traditional practices. Formulated at the level of a policy for socioeconomic development, its implementation was to seek out modernizing elites in the territories labeled as traditional, aiming to work through them to transform traditional societies into modern ones. In the era of decolonization, this approach came to be seen widely as "neocolonial" (a term developed in Africa in the 1960s), and it lost popularity.[4] Sociologist Norbert Elias took a distinctive position, arguing explicitly for group-level analysis of human behavior. His most original step was modeling a "Primal Contest," a model of a contest without rules, as the first step in a series of models of

[1] Von Bertalanffy, *General System Theory*.

[2] For a biographical review of Miller and his research, see G. A. Swanson, "James Grier Miller's Living Systems Analysis (LSA)," *Systems Research and Behavioral Science* 23 (2006): 263–71.

[3] John von Neumann and Oskar Morgenstern, *Theory of Games and Economic Behavior* (Princeton: Princeton University Press, 1944); John Nash, "Non-cooperative Games," *Annals of Mathematics* 54 (1951): 286–95. Von Neumann's initial paper on game theory appeared in 1928. For an extension of this reasoning to biology with the "evolutionarily stable strategy," see John Maynard Smith and G. R. Price, "The Logic of Animal Conflict," *Nature* 246, 5427 (1973): 15–18; Kurt Lewin, "Defining the 'Field at a Given Time,'" *Psychological Review* 50 (1943): 292–310; Lewin, "Frontiers in Group Dynamics: Concept, Method and Reality in Social Science; Social Equilibria and Social Change," *Human Relations* 1, 1 (1947): 5, 9–10. See also Lewin, "Frontiers in Group Dynamics II: Channels of Group Life; Social Planning and Action Research," *Human Relations* 1, 2 (1947): 143–53.

[4] Modernization theory categorized individuals and societies as "traditional" or "modern," distinguishing their readiness to innovate. In effect, this was a theory of human nature.

multi-person games.⁵ Elias emphasized interdependence of people, combining them into what he called "figurations," and using pronouns ("I," "we," "they") to convey their relative perspectives. In addition, he assumed that human social and biological characteristics must be tightly bound—he criticized Parsons sharply for treating them separately.⁶

Other systems-theorists sought to extend their framework to society and the history of civilizations, but again the oversimplified analysis gave few results. Jay Forrester at MIT, a founding figure in systems dynamics, published a 1971 model of the world based on the interaction of population and economy.⁷ Work on this project soon came to a halt for lack of funding and results; meanwhile, parallel studies in modeling climate managed to accelerate.⁸ A more lasting social analysis has been the World-Systems analysis formulated by Immanuel Wallerstein. This framework, developed for study of the rise of Atlantic capitalism, gave more attention to study of empirical data than to keeping up with new ideas in systems analysis.⁹ The framework has been applied to other historical situations and has had a substantial influence on historical, sociological, and anthropological literatures.

Physical sciences. New systems for dating the past continued to emerge. In laboratory work, carbon-14 was discovered in 1940 as a radioactive isotope with a half-life of 5700 years that exists in trace amounts in organic materials. By 1946, chemist Willard Libby had developed techniques for sampling and dating based on the ratio of carbon-14 and carbon-12. This technique was especially valuable for dating between 500 and 50,000 years ago, so that it became central to documenting human remains.[10] The International Geophysical Year (IGY) of 1958–1959, organized by the United Nations and the International Council for Science (ICSU), brought an unprecedented level of investment and international collaboration in geological study. This research led soon to verification of Wegener's 1912 theory of continental drift, the documentation of tectonic plates, and the study of expansion of the

---

[5] Norbert Elias, trans. Stephen Mennell and Grace Morrissey, *What Is Sociology?* (New York: Columbia University Press, 1978), 76–80. Elias modeled groups in qualitative discussion of figurations, rather than theorizing groups explicitly.

[6] "Man's biological endowments play no part in the formation of his social bonds ... if it is simply taken for granted—as it is by Talcott Parsons—that human personality structure is independent of social structure." Elias, *What Is Sociology?*, 134, citing Parsons, "Psychology and Sociology," in John Gillin, ed., *For a Science of Social Man* (New York: Macmillan, 1954), 84.

[7] Jay Forrester, *World Dynamics* (Cambridge, MA: Wright-Allen, 1971).

[8] For comparison of initial work in datasets and modeling for global society and climate, see Patrick Manning, *Big Data in History* (London: Palgrave Macmillan, 2013), 90–93. Climate analysis advanced in 1958 with measurement of atmospheric carbon dioxide. Spencer Weart, "The Discovery of Global Warming," http://www.aip.org/history/climate/index.htm.

[9] Immanuel Wallerstein, *The Modern World-System*, vol. 1 (New York: Academic Press, 1974).

[10] American Chemical Society National Historic Chemical Landmarks. Discovery of Radiocarbon Dating. http://www.acs.org/content/acs/en/education/whatischemistry/landmarks/radiocarbon-dating.html (accessed May 26, 2018).

ocean basin at mid-oceanic rifts. In addition, it was discovered that mid-oceanic expansion preserved the direction of magnetic polarity in newly formed oceanic floor, thus providing yet another way of dating the geological past. With these discoveries, the relevance of Milankovitch's theory of variations in insolation with shifts in the Earth's orbit became clear, and the cycles were initially traced back by 450,000 years.[11] Studies of climate change and climate modeling expanded, leading to the first report of the Intergovernmental Panel on Climate Change (IPCC) in 1990.

Another set of observations on geology and the Earth system came from quite a different direction. James Lovelock, a specialist in inorganic chemistry studying Mars with data from space probes, found that life was not likely to have formed on Mars, especially because of the great historic variations in Martian temperature. Lovelock compared Martian data to the new data that were emerging for the historical temperature of Earth and concluded that the temperature of Earth should have been much higher. To resolve this discrepancy, and in association with microbiologist Lynn Margulis, he developed a model in which biological species, especially single-celled algae that relied heavily on inorganic material, produced enough oxygen and carbon dioxide to modulate Earth's temperature. He labeled this phenomenon Gaia, after the Greek goddess of the Earth, arguing that Gaia's mechanisms of equilibration were robust enough that industrialization should not be seen as a threat to the environment.[12] The Gaia thesis on the overall functioning of the Earth system, proposed by a researcher based far outside of geology and using principles different from those of geology, was received with skepticism by geologists.

Biology. Biological sciences opened an era of remarkable innovation in the last half of the twentieth century, both in laboratory research and in public health applications.[13] In a 1943 experiment, Salvador Luria and Max Delbrück demonstrated that, in bacteria, genetic mutations arise randomly and without any process of genetic selection, rather than being a response to selection: this was one final step to confirm Darwin's vision of natural selection, in which mutations were random rather than determined by ecological need. In a further key step linking genetic theory to biology, Oswald Avery of Rockefeller University completed research in 1944 demonstrating that the chemical constituent carrying heredity was not that of proteins, as

---

[11] For application of Milankovitch's work, see A. Berger and M. F. Loutre, "Insolation Values for the Climate …" *Quaternary Science Reviews* 10 (1991): 297–317.

[12] Lovelock, James, *Gaia, A New Look at Life on Earth* (Oxford: Oxford University Press, 1979). James Lovelock and Sidney Epton, "The Quest for Gaia," *New Scientist* (February 8, 1975), 304.

[13] Patrick Manning, "The Life-Sciences, 1900–2000: Analysis and Social Welfare from Mendel and Koch to Biotech and Conservation," *Asian Review of World Histories* 6 (2018): 185–208; Patrick Manning and Mat Savelli, eds., *Global Transformation in the Life Sciences, 1945–1980* (Pittsburgh: University of Pittsburgh Press 2018).

had been widely thought, but was nucleic acids.[14] As discoveries developed in microbiology and what became known as molecular biology, a more detailed sense of the mechanism of genetic reproduction steadily emerged.[15] In particular, inheritance was not simply the passing on of characteristics generated by specific genes (as one might think from a basic Mendelian model). Instead, phenotypical characteristics were seen to result from complex mixes of genes, themselves undergoing recombination which further mixed the patterns of inheritance. The notion of genetic recombination was widely publicized through the 1946 work of Joshua Lederberg, a young researcher who devised inventive experiments with the *E. coli* bacterium that showed that its single strand of DNA nonetheless underwent recombination as it replicated, so that two genes that had previously been located near to one another on the chromosome might end up located a greater distance apart. Research then focused on learning the structure of DNA and its specific role in genetic reproduction. In 1953 work at Cambridge University, the American James Watson and British Francis Crick, supported by the X-ray crystallography of Rosalind Franklin, identified the double-helix structure of DNA molecules. In 1961, Marshall Nirenberg's group in the United States completed the first step in the validation of DNA code—linking DNA to RNA to an amino acid, phenylalanine. Work continued until the links for the 20 amino acids linked to DNA were established.[16]

Research on early stages in biological evolution brought important results. Lynn Margulis developed the notion of endosymbiosis—her 1970 book argued that creation of eukaryotes began when bacteria were absorbed into proto-eukaryotic cells, surviving as membrane-bound organelles, notably mitochondria. This was an important step in the genesis of eukaryotes, but many more coevolutionary processes were needed to achieve their ultimate structure—that of cells within genes on chromosomes within cellular nuclei

[14] The initial 1944 publication by Avery and his associates at the Rockefeller Institute was: O. T. Avery, C. M. MacLeod, and M. McCarty, "Studies on the Chemical Nature of the Substance Inducing Transformation of Pneumococcal Types: Induction of Transformation by a Desoxyribosenucleic [sic] Acid Fraction Isolated from Pneumococcus Type III," *Journal of Experimental Medicine* 79 (1944): 137–58.

[15] To put the relationships crudely: *Proteins are made up of a series of amino acids, linked in a chain. Nucleic acids, somewhat more complex, are made up of a chain of "nucleotides," each composed of a five-carbon sugar, a phosphate group, and a nitrogenous base. The two principal nucleic acids are DNA (deoxyribonucleic acid) and RNA (ribonucleic acid). DNA consists of two long strands in a double helix: the bases of its nucleotides are thymine, cytosine, adenine, and guanine; its sugar is deoxyribose. RNA molecules are shorter and have only one strand; the bases of its nucleotides are the same except that uracil replaces thymine; its sugar is ribose.*

[16] Joshua Lederberg and Norton D. Zinder, "Concentration of Biochemical Mutants of Bacteria with Penicillin," *Journal of the American Chemical Society* 70 (1948): 4267; J. D. Watson and F. H. C. Crick, "Molecular Structure of Nucleic Acids: A Structure for Deoxyribose Nucleic Acid," *Nature* 171 (1953): 737–38; Marshall Nirenberg and J. Heinrich Matthaei, "The Dependence of Cell-Free Protein Synthesis in E. Coli upon Naturally Occurring or Synthetic Polyribonucleotides," *PNAS* 47, 10 (1961): 1588–602.

that reproduced by mitosis. By 2010, this analysis led to an argument that eukaryotes had become well established by 850 million years ago. That date can be taken as the starting point for Darwinian biological evolution.[17]

Philosophical and epistemological reflections on these dramatic scientific advances arose in the 1970s, with attention especially to the issue of reductionism. Max Delbrück presented a set of 1974–1975 lectures in which he articulated the problems of "evolutionary epistemology."[18] Theodosius Dobzhansky and Francisco Jose Ayala (who had been his student) convened a conference on the philosophy of biology, publishing the results in 1974. Ayala emphasized the distinctions among three types of reductionism: ontological, methodological, and epistemological.[19] Yet the conference discussion was necessarily focused on past work and debate, so that there was little attention to a new reductionism that was just then arising.

A veritable explosion of new thinking burst into biological studies in the 1960s and 1970s, inspired by these and other advances.[20] In this knowledge-rich situation, biologists began to take a reductionist approach to social behavior, seeking to explain social change in biological terms. A newly triumphant molecular biology set off a scattershot of ideas in multiple directions, in contrast to the reductionism of the 1930s and its focus on genetic change.

---

[17] Lynn Margulis, *Origin of Eukaryotic Cells; Evidence and Research Implications for a Theory of the Origin and Evolution of Microbial, Plant, and Animal Cells on the Precambrian Earth* (New Haven: Yale University Press, 1970); Thomas Cavalier-Smith, "Origin of the Cell Nucleus, Mitosis and Sex: Roles of Intracellular Coevolution," *Biology Direct* 5 (2010): 1–78.

[18] Max Delbrück, *Mind from Matter? An Essay in Evolutionary Epistemology* (Oxford: Blackwell, 1986). The essays were transcribed by Delbrück's students, reviewed by Delbrück, and published after his death.

[19] Ayala's types of reductionism were: (1) ontological—the higher follows from the lower; (2) epistemological—molecular biology can explain other biological disciplines, e.g., genetics is explained by molecular biology; (3) methodological—efficiency in research as when physicists joined biology. The discussion also adopts the term "compositionism" to refer to the opposite of reductionism. Francisco Ayala, "Introduction," in Francisco Jose Ayala and Theodosius Dobzhansky, eds., *Studies in the Philosophy of Biology: Reduction and Related Problems* (London: Macmillan, 1974), viii–x.

[20] On punctuated equilibrium, see Niles Eldredge and S. J. Gould, "Punctuated Equilibria: An Alternative to Phyletic Gradualism," in T. J. M. Schopf, ed., *Models in Paleobiology* (San Francisco: Freeman Cooper, 1972), 82–115; republished in Stephen Jay Gould, *Structure of Evolutionary Theory* (Cambridge, MA: Belknap Press at Harvard University) and also as Gould, *Punctuated Equilibrium* (Cambridge, MA: Belknap Press at Harvard University, 2007). On population thinking, see Ernst Mayr, *Evolution and the Diversity of Life* (Cambridge, MA: Harvard University Press, 1975), 26–29. Mayr emphasized the centrality of population thinking in Darwin and in subsequent biological analysis. He was widely praised among biologists but his analysis nonetheless stopped short of observing the parallel advance of population thinking that had taken place in natural and social sciences. For attention to population thinking more broadly, consider the references herein to Dalton, Avogadro, Boltzmann, Curie, Rutherford, Fisher, and Maynard Smith: the issue is that of considering both the next smaller level and the next larger. For additional dimensions of this explosion in biological thinking, see Ernest Nagel, *The Structure of Science* (New York: Harcourt, Brace and World, 1961).

Population geneticists were early leaders in this new direction, exploring the "genetical evolution of social behavior." W. D. Hamilton, an associate of Ronald Fisher in Britain, published two closely reasoned theoretical articles in 1964 that posed possible conditions for biological evolution of social behavior.[21] In particular, Hamilton introduced the notion of "inclusive fitness," in which the genetic success of an individual is determined not only by the number of direct descendants but includes the descendants of siblings and other close relatives. This approach, since adopted widely, opened the door to exploration of cooperation and even altruistic behavior among individuals. The field of cladistics expanded to visualize relationships among various organisms or genetic characteristics. Willi Hennig, a German entomologist, developed new techniques for representing family trees; a 1966 translation of his work into English brought widespread adoption.[22] The field of ecology took a step forward as Paul R. Ehrlich and Peter H. Raven coined the phrase "coevolution" to describe the evolutionary interactions of plants and butterflies.[23] These and other initiatives assumed a focus on the level of the organism, meaning that any level of group or social activity is analyzed only at the individual level. These biological initiatives thus involved no discussion with those social scientists who sought to analyze group behavior in evolutionary terms. This choice was doubtless related, at some level, to the preponderance of individualistic philosophy which had established a firm grasp on intellectual life in the nineteenth century and which was reaffirmed in important ways in the late twentieth century.

Another mathematical approach to evolution arose from game theory: in 1972, John Maynard Smith applied game theory to Hamilton's 1964 work on the interaction of individual organisms and derived what has come to be known as an "evolutionarily stable strategy."[24] The emerging argument was

---

[21] Hamilton drew on revisions to the second edition of Fisher's pathbreaking 1930s volume. Ronald A. Fisher, *The Genetical Theory of Natural Selection*, 2nd ed. (New York: Dover, 1958); W. D. Hamilton, "The Genetical Evolution of Social Behaviour, I," *Journal of Theoretical Biology* 7 (1964): 1–16; Hamilton, "The Genetical Evolution of Social Behaviour, II," *Journal of Theoretical Biology* 7 (1964): 17–xx.

[22] Willi Hennig, trans. D. Davis and R. Zangerl, *Phylogenetic Systematics* (Urbana: University of Illinois Press, 1966 [1950]). The outcome of a cladistic analysis is a cladogram—a tree-shaped diagram (dendrogram) that is interpreted to represent the best hypothesis of phylogenetic relationships. Although traditionally such cladograms were generated largely on the basis of morphological characters and originally calculated by hand, genetic sequencing data and computational phylogenetics are now commonly used in phylogenetic analyses. On the issue of tree models vs. alternatives, see April McMahon, Thomas R. Trautmann, and Andrew Shryock, "Language," in Shryock and Smail, *Deep History*, 103–27.

[23] Paul R. Ehrlich and Peter H. Raven, "Butterflies and Plants: A Study in Coevolution," *Evolution* 18 (1964): 586–608.

[24] John Maynard Smith, "Game Theory and The Evolution of Fighting," *On Evolution* (Edinburgh: Edinburgh University Press, 1972); John Maynard Smith and George R. Price, "The Logic of Animal Conflict," *Nature* 246, 5427 (1973): 15–18; Nash, "Non-Cooperative Games." For later elaboration of the same insights, see Maynard Smith, *Evolution and the Theory of Games*.

that biological evolution of social behavior might emerge, but that the transmission of traits could only be through individual organisms and not through groups. Meanwhile, Luigi Luca Cavalli-Sforza, who had long been working on blood types as an indication of evolutionary patterns, teamed up with Australian mathematical biologist Marcus Feldman to publish a series of studies, beginning in 1973, on cultural and biological evolution.[25]

Epigenetics had just been added to the synthesis. The process of methylation of DNA, in which a certain hydrogen molecule on a nucleotide is replaced by a methyl group, was known even before the full structure of DNA was determined. Rather suddenly, by 1975, it became understood that methylation and other modifications of DNA could turn gene activity on or off, speed it up or slow it down, so that a given gene could have quite different effects on the phenotype of the individual organism, not just at conception and birth but during ontogeny or development. This new area of study became known as "epigenetics," and the results of epigenetics gave new insight into processes of embryology and ontogenic development. Stephen Jay Gould's 1977 synthesis showed that the development of individual organisms was governed not just by genetic DNA but by epigenetic proteins that regulated the activity of genes, changing the rate of growth and the shape of the organs or the entire body of an organism. In a parallel synthesis, François Jacob showed that "evolutionary tinkering" can occur also at the genetic level, with old genes being repurposed to serve new functions.[26]

To characterize the newly arising relationship between evolution and development, the term "evo-devo" was coined. Thus was born the combined study of phylogenic evolution and ontogenic development (evolution of whole species vs. the life-course development of individual organisms): this new framework resolved key questions by showing the flexibility of the combined system of DNA and epigenetics. Thereafter, "countless mechanisms of transmission" were documented, showing that changes brought by epigenetic transmission are a significant part of the phenotype of an animal.[27]

Other biologists pursued similar ideas in different ways, arguing that basic genetic change was sufficient to have brought about the full range of social change. E. O. Wilson advanced the broad project of "sociobiology" as a "new synthesis" in which social and biological evolution were seen as overlapping

---

[25] L. L. Cavalli-Sforza., and M. W. Feldman, "Models for Cultural Inheritance: I. Group Mean and Within Group Variation," *Theoretical Population Biology* 4 (1973): 42–55.

[26] Stephen Jay Gould, *Ontogeny and Phylogeny* (Cambridge, MA: Belknap Press of Harvard University Press, 1977); François Jacob, "Evolution and Tinkering," *Science* 196 (1977): 1161–66. The latter is a remarkably comprehensive yet concise statement of the current understanding of evolutionary processes.

[27] Gary Felsenfeld, "A Brief History of Epigenetics," *Cold Spring Harbor Perspectives in Biology* 6 (2014): PMC3941222; David Latchman, *Gene Regulation: A Eukaryotic Perspective*, 4th ed. (Cheltenham, UK: Nelson Thornes, 2002).

yet governed by biology.²⁸ Richard Dawkins published *The Selfish Gene*, arguing that the effect of biological evolution was to serve the needs of the gene, rather than the gene serving the needs of the organism. At the end of this book, he included an argument for a sort of cultural evolution, in which "memes" are invented as attractive expressions of tunes, phrases, or other devices, and that the memes develop in a "kinetic" process, a sort of natural selection to give birth to higher-order thinking. All of these biological scholars sought to extend biological evolution into developing social behavior.

Yet another expansion of biological thinking appeared in 1978 with psychologist James G. Miller's thick volume, *Living Systems*. In this detailed and rigorous application of systems thinking to biology, Miller identified living systems from the scale of the cell to supranational systems. Thus, he analyzed biological systems (from cell to organism), social systems (society and supranational system), and systems that were both biological and social (group and organization). He emphasized that living systems must perform many functions and, thus, are composed of relatively complex subsystems. Within each living system, he identified twenty subsystems for any living system, each performing a function necessary for the survival and reproduction of the system.²⁹ The study focused primarily on biological structure and functioning, but also included numerous observations on evolutionary change. This work had greatest resonance among psychologists; its detail and generality remain essential for study of large biological and social systems.³⁰

Paleontology in Africa brought important advances to evolutionary thinking. Most striking was the 1955 discovery in the Olduvai Gorge of Tanganyika, by Mary and Louis Leakey, of skeletal remains of a new hominid species, associated with pebble-sized stone tools. Louis Leakey announced the results in 1964, naming this relatively small bipedal species as *Homo habilis*, and labeling its tool-making industry as Oldowan.³¹ The announcement brought wide excitement and launched a debate about the definition

---

[28] Edward O. Wilson, *Sociobiology: The New Synthesis* (Cambridge, MA: Belknap Press of Harvard University Press, 1975). For another general argument, see Wilson, *On Human Nature*.

[29] Miller identified the parallels among all living systems, especially in their short-term functioning. But in a choice that I find surprising, he did not distinguish between biological and social mechanisms of change—that is, DNA-based biological evolution for cells, organs, organisms, and animal groups, in contrast to the social evolution of human communities, societies, and humanity overall. (He also neglected cultural evolution, though this new field had barely emerged by 1978.) In another surprising limitation, his functional approach gives little attention to conflict or disease in biological species or to war in humanity. If these deficiencies could be corrected, Miller's work would have wider relevance. James G. Miller, *Living Systems* (New York: McGraw-Hill, 1978); Swanson, "Miller's Living Systems Analysis," 267. Miller and Donald T. Campbell, though both held doctorates in psychology, appear not to have interacted.

[30] G. A. Swanson, "James Grier Miller's Living Systems Analysis (LSA)," *Systems Research and Behavioral Science* 23 (2006): 263–71.

[31] Leakey classified the new species within the genus *Homo*; South African archaeologist Raymond Dart proposed the species name *habilis* to refer to its able tool-making. L. S. B. Leakey, P. V. Tobias, and J. R. Napier, "A New Species of the Genus *Homo* from Olduvai Gorge," *Nature* 202 (1964): 7–9.

of the genus *Homo*.[32] In other paleontological work, as it became possible to date skeletons using radiocarbon methods, researchers working on Europe concluded that there was a "human revolution" about 40,000 years ago, as dramatic changes in tools, social practices, and in visual art made them conclude that a great transformation had occurred. (Later work showed that these advances had occurred earlier in Africa, and that the changes in Europe reflected the arrival of African *Homo sapiens* rather than innovation by Neanderthals born in Europe.)

In this same era, the United Nations had established the International Biological Program (IBP), 1964–1974, as a successor to the IGY. Initial plans for this collaborative research focused on the laboratory work of molecular biology, centered in a few industrial centers. A combination of representatives from ex-colonial nations and the Soviet bloc refocused research on broad concern for human welfare. The results did not bring such apparently spectacular theoretical advances as did IGY, yet the investment in broad collaboration and in social welfare led to the substantial strengthening of the emerging field of ecology, which was linked to growing social concerns about environmental degradation. Further, when the IBP is set in the context of the contemporaneous biological initiatives just listed, it can be argued that it is worth reviewing the relative neglect of its place in studies of history of science.[33]

Donald T. Campbell and Social Evolution. The explosion in biological thinking had significant reverberation in the field of psychology, where the wide-ranging Donald T. Campbell began publishing in 1965 on the value of Darwinian theory for study of sociocultural evolution.[34] Campbell delivered the 1975 presidential address of the American Psychological Association with a focus on sociocultural evolution that relied heavily on Darwinian thinking. His tour de force of theoretical and bibliographical comprehensiveness, summing up and expanding a decade of his analytical work, remains the best single statement of the case for systematic study of social evolution.[35]

---

[32] Leakey wished to extend the definition of genus *Homo* very broadly; others found that *habilis* was phenotypically closer to the preceding Australopithecines, arguing that *H. ergaster* was the beginning of the genus *Homo*. Tattersall, *Masters of the Planet*, 81–92.

[33] Manning, "Life Sciences, 1900–2000," 195–96. On the role of Soviet UNESCO officials in encouraging the IBP to focus on field studies of ecology rather than laboratory study of molecular biology, see Doubravka Olšakova, "From Sovietization to Global Soviet Engagement?" in Patrick Manning and Mat Savelli, eds., *Global Transformations in the Life Sciences, 1945–1980* (Pittsburgh: University of Pittsburgh Press, 2018), 99–113.

[34] Donald T. Campbell, "Variation and Selective Retention in Socio-Cultural Evolution," in Herbert R. Barringer, George I. Blandsten, and Raymond W. Mack, eds., *Social Change in Developing Areas* (Cambridge, MA: Shenkman Publishing Company, 1965), 19–49.

[35] Donald T. Campbell, "On the Conflicts Between Biological and Social Evolution and Between Psychology and Moral Tradition" (presidential address, American Psychological Association), *American Psychologist* (December 1975): 1103–26; reprinted as Donald T. Campbell, "On the Conflicts Between Biological and Social Evolution and Between Psychology and Moral Tradition," *Zygon* 11, 3 (1976): 167–208. See also Donald T. Campbell,

Despite the skill and erudition of his intervention, the minimal response to it showed how difficult it was to attempt a linkage of Darwinian evolution with social-scientific notions of change. He emphasized *variation, selective systems,* and *retention and duplication* as the three essential requirements of sociocultural evolution.[36] For instance, Campbell argued that a system of retention and duplication could function in a social setting:

> ... through social mechanisms of child socialization, reward and punishment, socially restricted learning opportunities, identification, imitation, emulation, indoctrination into trivial ideologies, language and linguistic meaning systems, conformity pressures, social authority systems, and the like, it seems reasonable to me that sufficient retention machinery exists for a social evolution of adaptive social belief systems and organizational principles to have taken place, in addition to the less problematic social evolution of technological devices ... [Yet] the mechanisms making possible such a social retention system would themselves be a product of biological and social evolution.[37]

Further,

> For a natural-selection type of socio-cultural evolution to work, the retention system must be capable of perpetuating uncomprehended functional recipes. The retention system must operate, as in biological evolution, by perpetuating everything it receives from the edited past. Inevitably this includes a lot of noise, maladaptive mutations, and chaff, along with the selected kernels of wisdom.[38]

Campbell extended his view of social evolution up to the present, with innovative arguments such as that the creation of more complex societies means that less of the innovations are adaptive. Thus, higher pressures for

---

"Evolutionary Epistemology," in P. A. Schilpp, ed., *The Philosophy of Karl R. Popper* (LaSalle, IL: Open Court, 1974), 412–63; Donald T. Campbell, "Unjustified Variation and Selective Retention in Scientific Discovery," 139–61; and Donald T. Campbell, "'Downward Causation' in Hierarchically Organised Biological Systems," 179–186, in Ayala and Dobzhansky, eds., *Studies in the Philosophy of Biology*. In the same volume, see also Theodosius Dobzhansky, "Chance and Creativity in Evolution," 307–38; and Francisco J. Ayala, "The Concept of Biological Progress," 339–55.

[36] Phrasing it slightly differently in the same address, Campbell said, "The evolutionary theory I employ is a hard- line neo-Darwinian one for both biological and social evolution, the slogan being 'blind-variation and systematic selective retention.'" Campbell, "Biological and Social Evolution," 1104.

[37] Campbell, "Biological and Social Evolution," 1107.

[38] "We psychologists should also be more sympathetic to a socially useful age-specific role, if one exists, for women past child-bearing age to become tradition-enforcing, moralizing scolds, instead of interpreting this as a dysfunctional, neurotic reaction formation." Campbell, "Biological and Social Evolution," 1108.

conformity, long apprenticeships, and castes arose until the industrial revolution. Thereafter, he argued, technological complexity has been accompanied by reduced conformity pressure and greater freedom to change jobs—so that the cumulated technological wisdom is now embodied in industrial machines rather than in humans.

Campbell noted critically the contemporary efforts of biologists to treat social evolution as an "unproblematic extension of the biological process."[39] He launched a long and forceful critique of biological arguments against the possibility of groups as having any evolutionary functionality.[40] He concluded with two hypotheses:

1. Human urban social complexity has been made possible by social evolution rather than biological evolution.
2. This social evolution has had to counter individual selfish tendencies which biological evolution has continued to select as a result of the genetic competition among the cooperators.[41]

Campbell's analysis, while outstanding for its day, would need to be updated to be applied today. For instance, he treated modern human heritage as having descended entirely from great urban societies, taking insufficient account of the amount of human heritage carried forth in small and rural communities. His vision of the timing of social evolution was that it began with urbanization, some five thousand years ago. In addition, and despite his critique, the theorists of "cultural evolution" have since created a large literature based on the logic of individual-level learning. Thus, any analyst today who chooses to emphasize the role of social evolution—as I do—must either accept that there exist three categories of human evolution (biological, cultural, and social) or devote energy to challenging the existence of cultural evolution by seeking to demonstrate that all human evolutionary change can

---

[39] "There is a parallel tendency for biologists (mainly ethologists) interested in social systems and morality to present evolutionary arguments in which social evolution is mentioned as an incidental unproblematic extension of the biological process." He cites Lorenz, Teilhard de Chardin, and Wilson, *Sociobiology*. Campbell, "Biological and Social Evolution," 1108.

[40] "J. B. S. Haldane (1932) devoted an appendix to the barrier against selecting characteristics that are adaptive for the group but costly to the individual, in species in which there is genetic competition among the cooperators. He initiated the use of the term *altruism* for such characteristics (meaning *self-sacrificial* altruism) and started the talk about genes for altruism and what would happen to them in the course of natural selection. This argument lay neglected until the excessive claims of Wynne-Edwards (1963) for group selection led it to be revived." Campbell, "Biological and Social Evolution," 1111; Vero Copner Wynne-Edwards, *Animal Dispersion in Relation to Social Behavior* (London: Oliver & Boyd, 1962); George C. Williams, *Adaptation and Natural Selection: A Critique of Some Current Evolutionary Thought* (Princeton: Princeton University Press, 1966), 97, 110–22.

[41] Campbell, "Biological and Social Evolution," 1115.

be classified either as biological evolution at an individual level or as social evolution at a group level.[42]

Social Sciences. Social scientists pursued their interest in patterns of change and innovation at the level of social groups. This emphasis put them in touch with studies in human paleontology but did not otherwise put them in touch with biological studies.[43] Certain anthropologists, in the mid-twentieth century, returned explicitly to the study of group-level social and cultural evolution. V. Gordon Childe, Australian-born and an English-trained archaeologist, focused on Europe and West Asia and developed a vision of stages of advancement focusing on the agricultural revolution and the urban revolution, both occurring in the same regions of the eastern Mediterranean and the Fertile Crescent.[44] Leslie White envisioned himself as a successor to L. H. Morgan: he rejected the stance of Boas as anti-evolutionary. White focused not on cultural interaction but on the technological development within any society, as measured by the quantity of energy consumed.[45] Julian Steward contested both the relativist stance of Boas and the stage theories of White, seeking a "multilineal" middle ground in social evolution.[46] George Peter Murdock, from an empiricist standpoint, led in building an immense database of observations on ethnic groups worldwide, assembled from an overview of ethnographic reports: this dataset became the Human Relations Area Files (HRAF), based at Yale University.[47] African studies were expanding significantly in this era, and Murdock, though not a specialist in Africa, analyzed the African data in HRAF and published a 1959 book including numerous hypotheses on culture history, classifying ethnicities through material culture.

---

[42] On cultural evolution, see Maynard Smith and Price, "The Logic of Animal Conflict"; Boyd and Richerson, *Culture and the Evolutionary Process*, 227–40. As will be seen, I am taking the position of affirming three types of evolution, while Christopher Hallpike has taken the position of denying the existence of cultural evolution. Hallpike, *How We Got Here: From Bows and Arrows to the Space Age* (Central Milton Keynes, UK: AuthorHouse, 2008).

[43] Social scientists in many different fields contributed to an expanding—if not yet interconnected—discourse with implications for human evolution. Examples included the work of Leslie White and Robert Carneiro in anthropology, Donald T. Campbell in psychology, and Immanuel Wallerstein in sociology.

[44] Childe provided effective summaries of the differences among the evolutionary visions of Darwin, Spencer, Tylor, Morgan, diffusionist and functionalist anthropologists, and placed hope in the archaeological record. Childe, *Social Evolution* (London: Watts & Co., 1951), 3–14.

[45] "Culture must be explained in terms of culture. Thus, paradoxical though it may seem, 'the proper study of mankind' turns out to be not Man, after all, but Culture." Leslie White, *The Science of Culture: A Study of Man and Civilization* (New York: Farrar, Straus and Company, 1949), 141. Leslie White, *The Evolution of Culture: The Development of Civilization to the Fall of Rome* (New York: McGraw-Hill, 1959).

[46] Julian Steward, *Theory of Culture Change: The Methodology of Multilinear Evolution* (Urbana, IL: University of Illinois Press, 1955).

[47] Works based on HRAF were published by Robert Carneiro on development of states, Orlando Patterson on slavery, Frederic Pryor on the origin of the economy, and Peter Turchin on state development.

Bruce Trigger, in his review of sociocultural anthropology, used the term "American neo-evolutionism" to describe White and those who were close to him, emphasizing that their views emphasized ecological, demographic, and technological determinism, abandoning individual creativity, and explaining behavior based on constraints rather than free will and rationality.[48] Robert Carneiro, a student of Leslie White, relied on the Murdock data and other data to conduct statistical tests that suggested a correlation of increasing population density and higher degrees of social and political hierarchy.[49] These were efforts to present orderly patterns in social evolution—at the phenotypical level—without articulating underlying mechanisms to support them.

This revival of evolutionism in anthropology, under the leadership of White and Childe, had become explicit in its assumptions. In response to the accumulation of data in anthropology and archaeology, White and Childe chose to assert that social evolution began with the rise of agriculture in the Holocene epoch—crudely, 10,000 years ago. They emphasized a phenotypical approach in that they emphasized overall societal outlines, including general statements on agricultural and other technology. Murdock's 1959 study of Africa was comprehensive in description of social groups throughout the continent, yet his main emphasis was on his confirmation of several centers of agricultural innovation and his search for matrilineal social organization.[50] Overall, White's approach to social evolution gave prime attention to agricultural innovation and left non-agricultural societies out of consideration. The main mechanism suggested for social change was the gradual accumulation of surplus output in agriculture. In effect, White suggested that some societies evolved and others did not.[51]

The field of linguistics emerged, in the 1950s, from a long period of avoidance of the origin of language—a period initiated by the famous 1866 decision of the Linguistic Society of Paris to ban discussion of the origin of language, because the debates seemed so inconclusive.[52] New initiatives in

---

[48] Trigger centers this group on White and includes Julian Steward, Marshall Sahlins, Elman Service, Marvin Harris, Robert Carneiro, and Kent Flannery. Bruce Trigger, *Sociocultural Evolution: Calculation and Contingency* (Oxford: Blackwell, 1998), 124–51.

[49] Robert L. Carneiro, "Scale Analysis, Evolutionary Sequences, and the Rating of Cultures," in R. Naroll and R. Cohen, eds., *A Handbook of Method in Cultural Anthropology* (Garden City, NY: Doubleday, 1970), 846; Miller, *Living Systems*, 747, 756–760.

[50] George Peter Murdock, *Africa: Its Peoples and their Culture History* (New York: McGraw Hill: 1959).

[51] In a linkage of psychology and biology, during the 1960s and 1970s, psychologist Donald T. Campbell developed his case for the logic of social evolution. Yet no matter how powerfully Campbell was able to argue for the plausibility of social evolution, he was not able to articulate a specific mechanism by which such evolution would take place. Campbell's view of social evolution assumed that the process was brought into existence by urban life—that is, with what Childe had called the "Urban Revolution."

[52] A leading linguist who conducted work in the interim, cited widely in the late twentieth century, was Ferdinand de Saussure, trans. Wade Baskin, *Course in General Linguistics* (New York: McGraw Hill, 1966 [1916]).

linguistics during the 1950s came especially from Morris Swadesh, Joseph Greenberg, and Noam Chomsky.[53] Greenberg began his career in linguistics with a detailed classification of African languages into four major phyla or families.[54] Greenberg's classification faced one great debate, on the origin and spread of the Bantu languages of central and southern Africa. In debate with other scholars, Greenberg was able to show that the homeland from which proto-Bantu spread was not at the geographic center of the languages but at the northwest fringe of the distribution, next to related non-Bantu languages. This debate confirmed the methods and results of Greenberg's approach and strengthened African studies.[55] Having analyzed hundreds of African languages, Greenberg saw the next logical step as making the case for language universals—similarities across languages in syntax, morphology, and typology—in a 1963 article, a 1966 book, and in a four-volume, co-edited collection that appeared in 1978.[56] Chomsky, in quite a different approach, challenged the focus on language communities, focusing instead on questions about individual capability for speech, especially learning by children. He argued that his notion of universal grammar, inherent in all humans, meant that its study through English was sufficient. He was concerned less with the making of sounds than with the logic of constructing sentences and complete thoughts. He initially proposed that there must be an organ located within the brain that provided speech capability, but the search provided no results. Debates opened by Chomsky on the origins of language continued

[53] Morris Swadesh, a specialist in historical linguistics of North America, proposed in 1952 a system of "glottochronology," intended to locate the chronological time at which languages had first separated. Tests of this method showed that its precision was not high: rather than persist in revising the technique, most linguists retreated to estimating the ordinal separation among languages and gave up on estimating the chronological differences among them. Working with a list of 200 basic words assumed to exist in all languages (later reduced to a 100-word list), Swadesh assumed a constant rate of change of words at a rate of roughly 14% per millennium, and with this technique proposed estimation of temporal differences among languages. Morris Swadesh, *The Origin and Diversification of Languages* (London: Routledge & Kegan Paul, 1972), including a chapter entitled, "What Is Glottochronology?" 271–84. See also Joseph Greenberg's later theorization of glottochronology: Greenberg, "A generalization of glottochronology to *n* languages" (1987), reprinted in Croft, *Genetic Linguistics*, 108–13.

[54] Greenberg's work revised earlier classifications of African languages, notably by Dietrich Westermann. Joseph H. Greenberg, *The Languages of Africa* (Bloomington, IN: Indiana University, 1963); William Croft, *Joseph Harold Greenberg, 1915–2001: A Biographical Memoir* (Washington, DC: National Academy of Sciences, 2007).

[55] Joseph H. Greenberg, "Linguistic Evidence Regarding Bantu Origins," *Journal of African History* 13 (1972): 189–216.

[56] "The fundamental observation of *Language universals* is that pairs of linguistic categories in phonology, grammar and the lexicon typically show asymmetrical behavior that is to a very large extent cross-linguistically uniform." Martin Haspelmath, "Preface," in Joseph H. Greenberg, ed., *Language Universals, with Special Reference to Feature Hierarchies* (Berlin: Mouton de Gruyter, 2005 [1966]), vii; Joseph H. Greenberg, Charles A. Ferguson, and Edith A. Moravcsik, eds. *Universals of Human Language*, 4 vols. (Stanford: Stanford University Press, 1978).

for decades and became linked to questions of human nature.[57] Similarly, Greenberg's initiatives on language classification and on the study of universal patterns in language brought decades of research and debate.

Postwar economic analysis focused heavily on developing the macroeconomics articulated by John Maynard Keynes, emphasizing ways that government revenue and expenditure could influence employment levels, act to offset business cycles, and encourage steady growth of national economies. Yet new, microeconomic thinking gave increasing attention to individual firms and consumers: Ronald Coase published influential papers on transactions costs and the nature of social cost. Friedrich Hayek challenged Keynes with a vision of the self-regulating market as the complete social system. Hayek, who had done some work in biology, argued that the price mechanism in human society was a precise parallel to natural selection as the mechanism of change in organic life.[58] On this basis, he opposed any sort of central planning as fumbling and incomplete; he ultimately argued that efforts to address inequality or injustice do more harm than good. Hayek's annual meetings carried forth his campaign for individualism, identifying the market as the essential human institution.[59] Economist Milton Friedman, an analyst of monetary economics (as Keynes and Hayek had been), became an energetic advocate of Hayek's philosophy, applying it in a widely publicized economic development program in Chile under the military regime that had come to power in a 1973 coup d'état.

The debate was complex and is worth revisiting. Economist Mancur Olson analyzed contemporary behavior of social groups, documenting cases in which apparent group thinking reduced to individualistic logic; his critique became important in campaigns to limit US trade unions.[60] In the debate over whether group behavior was logically feasible or whether it inevitably reduced to individual-level behavior, Olson (in 1965, for economics, expanding on Hayek) and Williams (in 1966, for biology) each showed the flaws in naïve versions of group-behavior theses. Those who supported the thesis of distinctive group behavior (for instance, Lewontin in both biological and social change, as well as Service and Campbell in social change) made no detailed response to the critics. Supporters of group behavior continued to

[57] Noam Chomsky, *Syntactic Structures* ('s-Gravenhage: Mouton, 1957); Noam Chomsky and Michel Foucault, *The Chomsky-Foucault Debate on Human Nature* (New York: The New Press, 2006).

[58] Hayek's formulation provided a basis for rejecting all group thinking, especially among economists. It led later to rational choice theory and its assumptions in analysis and ideology. The key text is Hayek's paper, "Scientism and the Study of Society," serialized in *Economica* (1942–1944), and later published as the first part of Hayek, *The Counter-Revolution of Science* (New York: Free Press, 1955).

[59] Hayek's vision went beyond the general-equilibrium economic analysis of Walras and Pareto to propose markets and prices as the essence of the full social system. Friedrich A. Hayek, *Studies in Philosophy, Politics and Economics* (Chicago: University of Chicago Press, 1967).

[60] Olson, *The Logic of Collective Action*.

believe that it had an underlying logic but offered no specifics to articulate the mechanism. For lack of a firm defense, the logic of a distinctive group behavior in humans dissipated in academic life and in social practice.

In the postwar era, the discipline of political science expanded its consideration of contemporary politics at the global level and thereby opened the door to longer-term studies of relations among polities. Hans Morgenthau published *Politics among Nations* in 1948. This initial text on international relations might have been expected to reflect the contemporary wave of decolonization; it focused in fact on great-power struggles from the era of World War I.[61] Nevertheless, from this basis, the field of international relations within political science expanded its scope dramatically in the decades thereafter.

The discipline of history, in the postwar era, moved cautiously toward considering the era before 1500 but had little to say on evolution. The Annales tradition produced a memorable volume with the publication of Fernand Braudel's multidisciplinary exploration of the Mediterranean world in the sixteenth century. Maurice Halbwachs developed the notion of collective memory, distinguishing the consciousness of a community about its past from the documented writings of historians: this distinction gained steadily wider attention, though the subjective nature of memory had to be acknowledged. Social history and economic history each developed as much stronger subfields within history. Economic history had both its quantitative-theoretical (cliometrics) and its narrative (socioeconomic) versions. Social history too had its quantitative microhistory dimension and a qualitative version. The best-known example of the latter was *The Making of the English Working Class*, by E. P. Thompson, which emphasized Marxian versions of class consciousness and the agency of the workers who faced restrictions in the developing capitalist order. Thompson's analysis focused on working-class agency in contrast to elite initiative.[62] In a quite separate but influential appendage to historical studies, psychiatrist and amateur demographer Colin McEvedy published a dozen historical atlases from 1961 to 1984, beginning with medieval history and expanding to world history—the population estimates in these volumes—over long time periods and at regional, national, and civilizational levels—were relied upon for decades thereafter.[63]

World history announced itself as a field with the publication of *The Rise of the West*, in which William H. McNeill reviewed the interactions of

---

[61] Hans Morgenthau, *Politics Among Nations: The Struggle for Power and Peace* (New York: Alfred A. Knopf, 1948).

[62] Fernand Braudel, trans. Siàn Reynolds, *The Mediterranean and the Mediterranean world in the age of Philip II*, 2 vols. (New York: Harper and Row, 1976); Maurice Halbwachs, ed. and trans. Lewis A. Coser, *On Collective Memory* (Chicago: University of Chicago Press, 1992 [1950]); E. P. Thompson, *The Making of the English Working Class* (New York: Pantheon, 1964).

[63] Colin McEvedy and Richard Jones, *Atlas of World Population History* (Harmondsworth, UK: Penguin, 1978).

civilizations, religions, and social orders over five millennia, up to the year of its publication in 1963. This can be seen as another example of a global interpretation of history following a great war.[64] At the same time, the growth in area-studies academic programs indicated that Africa, South Asia, West Asia, and Latin America would be given serious attention within historical studies: regional histories increased in number, and interactions among these regional histories began to be considered. Studies of disease, environmental change, and cross-cultural trade were among the topics through which world-historical analysis expanded its scope. Slowly, grounds were being established for global analysis of the past.[65]

At the level of public policy, the Club of Rome formed in 1968, under the leadership of Aurelia Peccei, an Italian industrial manager, to develop a global vision on the "world problématique." In 1970, the group met at MIT with Jay Forrester, whose global socioeconomic model addressed many similar issues. The result was publication in 1972 of *The Limits of Growth*, which reported on results of Forrester's World 3 model and made policy recommendations. The book became widely popular and its general interpretation held up well over time, as shown in a 2004 update to the volume, but the elementary nature of the Forrester model could not rapidly be overcome.[66] As a result, attempts to model global society halted for some time and came to be considered impractical. Research on global society drew on a narrow range of domain experts, faced complex analytical questions, and did not succeed in expanding its funding, especially because of skepticism about its arguments that growth in various areas needed to be limited.[67] One must note the contrast between this research on global society and the research on global climate, which began just a few years earlier. Research on climate, while it began on a very small scale, managed to accelerate in its analysis and

---

[64] Historians wrote global interpretations in the wake of the Napoleonic wars and after World War I; Karl Jaspers followed suit after World War II. McNeill not only wrote a global interpretation but gave leadership thereafter in the formation of an organized field of world historical studies. William H. McNeill, *The Rise of the West: A History of the Human Community* (Chicago: University of Chicago Press, 1963).

[65] For a longer-term view of the exploration of world history in the US, at three major universities and with support of the Ford, Carnegie, and Rockefeller foundations, see Katja Naumann, *Laboratorien der Weltgeschichtsschreibung: Lehre and Forschung an den Universitäten Chicago, Columbia und Harvard 1918 bis 1968* (Göttingen: Vandenhoeck & Ruprecht, 2019).

[66] Forrester, *World Dynamics*; Donnella H. Meadows, Dennis L. Meadows, Jørgen Randers, and William W. Behrens II, *The Limits to Growth: A Report for the Club of Rome's Project on the Predicament of Mankind* (New York: Universe Books, 1972). See also Donnella H. Meadows, Jorgen Randers, and Dennis Meadows, *The Limits to Growth: The 30-Year Update* (White River Junction, VT: Chelsea Green Publishing, 2004).

[67] In addition, the macro-level focus of the Forrester model did not account sufficiently for the many levels of human society; historical and contemporary data were neither readily available nor consistently coded. Similar problems limited later efforts to prepare global historical datasets. Patrick Manning, *Big Data in History* (Houndmills, Basingstoke, UK: Palgrave Macmillan, 2013), 90–107.

its funding, and by the late 1980s had developed immense datasets and clear interpretations.[68]

Two new frames of analysis. In a response to the social movements of 1968, Immanuel Wallerstein—a professor of sociology and specialist in African studies at Columbia University who became an activist and a chronicler in the struggles at the university—shifted his research agenda and expressed his concerns about capitalism through an innovative analysis of the modern World-System.[69] This expansion on center-periphery analysis set a European-centered World-System as the unit of analysis, but focused on interactions and evolution in the three subgroups of center, semi-periphery, and periphery, allowing as well for external regions; the framework drew substantially on Annaliste and Marxian analysis. At Binghamton University, Wallerstein and sociologist Terence Hopkins formed the Fernand Braudel Center in 1976 to support studies within this paradigm.[70]

Also in the 1970s arose the new field of study now known as "cultural evolution," initially known as "cultural inheritance" or "dual inheritance." Two California-based groups launched this field, each modeling choices of individuals that had implications for their level of learning, arguing that the accumulation of these choices conveyed a lasting cultural inheritance. At Stanford, Luigi Luca Cavalli-Sforza, a well-known scholar who had long explored human evolution through study of blood types, joined with Marcus Feldman, a young Australian statistician, to work on modeling the transmission of cultural characteristics, publishing their first article in 1973.[71] At the University of California–Davis, ecologists Peter J. Richerson and Robert Boyd began work on what they called a "dual inheritance" of biological and micro-cultural paths.[72] The first publication of Boyd and Richerson was in a special journal issue responding to Donald Campbell's 1975 presidential address. In this article and in a longer version of it that appeared subsequently, Boyd

---

[68] While climate modeling became very successful, efforts, to implement its policy implications encountered the same resistance to change as global social modeling. James Hansen, Storms of My Grandchildren: The Truth about the Coming Climate Catastrophe and Our Last Chance to Save Humanity (London: Bloomsbury, 2009).

[69] Wallerstein, *The Modern World-System*, Vol. 1 (New York: Academic Press, 1974). Subsequent volumes were Vol. 2 (1980), Vol. 3 (1989), and Vol. 4 (2011). See also Immanuel Wallerstein and Paul Starr, eds., *The University Crisis Reader* (New York, Random House, 1971).

[70] The Fernand Braudel Center for the Study of Economies, Historical Systems, and Civilizations remained a major research center until its closure in 2020. Wallerstein directed it from 1976 to 2005. Among its leading associates were Giovanni Arrighi, Samir Amin, Andre Gunder Frank, and Beverly Silver.

[71] Cavalli-Sforza and Feldman, "Models for Cultural Inheritance"; L. L. Cavalli-Sforza and M. W. Feldman, "Towards a Theory of Cultural Evolution," *Interdisciplinary Science Reviews* 3 (1978): 99–107; L. L. Cavalli-Sforza and M. W. Feldman, *Cultural Transmission and Evolution: A Quantitative Approach* (Princeton: Princeton University Press, 1981).

[72] Richerson's PhD was in zoology but he moved into environmental studies and ecology; Boyd completed his PhD in ecology with Richerson.

and Richerson gave praise to Campbell for his emphasis on Darwinian criteria in considering sociocultural evolution but chose to bypass and almost totally neglect Campbell's emphasis on social evolution through the actions of groups.[73] The work on cultural inheritance by the two new groups was at the frontier of biology and social science. Cavalli-Sforza and Feldman identified themselves mainly with biology, while Boyd and Richerson identified themselves primarily with social science, as indicated by the journals in which they published.

Understandings of evolution by 1980. The postwar era, in which Cold War confrontation threatened nuclear war and "mutually assured destruction," was also a time of massive investment in social welfare. Competing social orders committed unprecedented investments, so that levels of public health and education rose rapidly in the many decolonizing nations, in the expanding number of nations ruled by communist parties, in war-torn Japan and Germany, and in the United States and its allies. In this atmosphere, impressive advances unfolded in many areas of research. Big changes in biological studies brought definitive discoveries, new fields of analysis, and fundamental debates. Social sciences research expanded in its spatial and temporal scope and in its methodological sophistication. Varying and often contradictory interpretations vied for attention in analysis of social change, growth, progress, and diversity. Investment in Earth sciences brought a major reorganization and expansion of the field, paralleled by the rise of environmental studies. Further, new fields arose in cybernetics, game theory, systems analysis, and advanced computation—they remained distinct but were eventually to coalesce under the umbrella of information sciences.

The neo-Darwinian theory or modern synthesis, combining Darwinian natural selection and Mendelian genetics, now expanded to include the results of molecular biology. Recombination of chromosomes in reproduction, increasing the diversity in the genome, came to be understood. DNA became known as the repository of genetic information, producing RNA and then proteins that carried on the functions of cells and organisms. Study of methylation in DNA opened up the field of epigenetics and the regulation of genes. Application of systems thinking to biology created the comprehensive framework of living systems. Such expanded knowledge of evolutionary

---

[73] Campbell's essay, they argued, "interprets the findings of sociobiology as precluding the existence of complex altruistic human societies and hypothesizes that group-level selection of culture opposes the individual selfishness favored by selection on genes. The main difficulty with this interpretation is that it leaps far past present theoretical and empirical understanding of dual inheritance systems." Robert Boyd and Peter J. Richerson, "A Simple Dual Inheritance Model of the Conflict Between Social and Biological Evolution," *Zygon, Journal of Religion and Science* 11 (1976): 129–31. This is one of nine commentaries on Campbell's 1975 essay, also reprinted in this issue. See also Peter J. Richerson and Robert Boyd, "A Dual Inheritance Model of the Human Evolutionary Process I: Basic Postulates and a Simple Model," *Journal of Social and Biological Structure* 1 (1978): 127–54; this article cites and briefly comments on the parallel work of Cavalli-Sforza and Feldman.

processes brought efforts to expand the logic in other directions, as by seeking ancient human remains. For the Pliocene era, work by paleontologists—especially the Leakeys—began a stream of discoveries. For the Pleistocene, work divided topically into paleontology, notably on Neanderthals, and studies of technology.

The United Nations supported two transformative international research efforts, as well as a successful campaign to eradicate smallpox. The 1957–1958 International Geophysical Year facilitated the clarification of plate tectonics theory and other new knowledge in oceanography. The 1964–1974 International Biological Program supported widespread research in ecology, enabling that field to establish its reliance on systems analysis. Other work in the physical sciences developed carbon-14 dating, followed by other dating techniques relying on radiometry, isotope ratios, magnetic polarity, and luminescence.[74]

Research in the social sciences, while it was short in definitive discoveries, expanded greatly in data collection and methodology: geographically with the rise of area studies, quantitatively with the application of statistical methods, and with extension into additional disciplines. Linguistics opened major new questions of micro-level origins and macro-level universals of language. Psychology explored language and behavior at individual and group levels. Anthropology expanded the work of correlating what had been dispersed datasets. Historians expanded the spatial, temporal, topical, and evidentiary base for their investigations.

In an initial wave of social science research, V. Gordon Childe labeled an Agricultural Revolution (ca. 10,000 years ago) and an Urban Revolution (ca. 5000 years ago), setting a framework that continues to be central in social science analysis. Anthropologist Leslie White posed a formal paradigm of social evolution, taking the Urban Revolution as the moment of its initiation and tracing levels of energy consumption as measures of growth and evolution. (White implicitly limited the beneficiaries of evolution to agriculturists.) In a contrast, Joseph Greenberg's campaign of language classification for African and other regions, later followed by analysis of language universals, gave indications that languages and speaking communities had begun well before the rise of agriculture. G. P. Murdock's global ethnographic database gave attention to agriculture but also showed that social structures had formed earlier.

In schools around the world, the postwar era brought creation of history curricula intended to convey a global understanding.[75] Periodization of these curricula, both in ex-colonies and in metropolitan regions, focused first on the era of capitalist empire but also on the era of civilization, especially Western Civilization. The curricula included brief introductory references to

---

[74] On luminescence, see Chapter 3.

[75] Shingo Minamizuka, ed., *World History Teaching in Asia* (Great Barrington, MA: Berkshire Publishing, 2019).

the rise of agriculture. William McNeill's *Rise of the West* gave recognition to a global-civilizational framework in the discipline of history.[76] With time, however, decolonization and critique of racial hierarchy tended to undermine notions of civilizational hierarchy.

A second wave of social science research arose along with the wave of global social protest that centered on the year 1968, including approaches that ranged from the politically radical to the conservative. New publication of works by Marx and Weber had widespread influence.[77] The field of anthropology, accused of complicity with colonialism, underwent a severe self-critique; a new focus on human agency arose in studies of social and economic history; efforts at global analysis arose in history and sociology; and economists developed a critique of group analysis and government regulation. These initiatives, while energetic, remained constrained within disciplines and sub-disciplines. Especially within social sciences but also within biological studies, as the volume of research expanded, evolutionary studies of hominin species remained segregated into temporal, spatial, and topical subfields.

The era from 1945 to 1980 was perhaps the high point of the disciplinary organization of knowledge. Yet in that same era, investigation of numerous issues began to require assembling knowledge from several disciplines. In this regard, researchers who focused especially on cross-disciplinary research and writing are of particular interest. Thus, psychologist Donald Campbell, psychologist James G. Miller, and geneticist L. L. Cavalli-Sforza led initiatives that led less to verifiable results than to inspiration and further inquiry. Nevertheless, it is difficult to conceive of today's level of cross-disciplinary research on issues in human evolution without considering the energy and imagination of their explorations. Meanwhile, out of the debates of the 1960s and 1970s arose two fields of study, cross-disciplinary at base, that were to be influential in subsequent times: cultural evolution, relying on advances in biology to seek explanations of how biological processes might have generated social structures; and World-Systems theory, tracing interaction of large-scale social units, first in the early modern era and then at longer time scales.

---

[76] McNeill, *Rise of the West*.

[77] Max Weber, Guenther Roth, and Claus Wittich, eds., *Economy and Society: An Outline of Interpretive Sociology*, 3 vols. (New York: Bedminster Press, 1968); Karl Marx, ed. and trans. David McLellan, *The Grundrisse* (New York: Harper and Row, 1971).

CHAPTER 11

Ecology and Society in a Neoliberal Era, 1980–2010

While there was no sharp turning point between periods, the era from 1980 to 2010 turned out to be sharply different from the preceding years of 1945–1980; it was commonly referred to as an era of "globalization." Trade union movements worldwide, having been at peak power, lost their influence rapidly. Neoliberal ideology took form and brought widespread reduction in state regulation of private enterprise. Public and university education continued to expand, but costs rose rapidly and the student body was drawn increasingly from wealthy families. A huge wave of social movements in 1989–1990, mostly opposing governmental restrictions, brought down numerous governments: especially in Africa and most spectacularly in socialist-led countries. The 1992 collapse of the Soviet Union shifted the world order, bringing another wave of decolonization. Public enterprises in socialist countries were sold off to buyers, so that Russia became fully capitalist. China, expanding economically since 1978, became by 2000 a capitalist economy governed by a communist party. The European Union, formed in 1993, expanded almost immediately to 27 member states. Throughout this period, economic inequality grew rapidly, especially within nations. In the academic and corporate worlds, this was a time of rapid expansion in computing technology and in the scale of data analysis. Creation of the Internet expanded electronic communication at every level of human existence. Epigenetics and evo-devo led to many innovative theories in biological affairs. Ecology expanded its system-based analysis. The Intergovernmental Panel on Climate Change (IPCC) formed in 1988 and made its initial report in 1990, with a critical review of climate change.

By 1980, work in the natural sciences had achieved impressive maturity in theory and empirical research, in laboratories and field studies. Physics and chemistry were perhaps most advanced, with over two centuries of intensive analysis. Geology, also with a long and effective discourse, had just achieved impressive advances. And biology achieved a succession of remarkable advances in the mid-twentieth century with the breaking of the DNA code followed by the recognition of epigenetics. Perhaps unexpectedly, the pace of scientific discovery accelerated even further in the years after 1980, so that an extraordinary increase in available data and in research resources has since created an intensive, new scientific dynamic in field after field. Any serious study of change in human society must give close attention to the new developments in biological and physical sciences—though without neglecting their emergence out of principles established in earlier work. For instance, multi-scale analysis became a formal expectation in all fields, while links across disciplinary lines were increasingly common.

For the social sciences by 1980, the problems were quite distinctive and the resources for study were more limited. The great variety in human institutions across the world's spaces, the rapid change in human social order and technology, combined with the limitations and unevenness of documentation, meant that the social sciences were far from achieving the sweeping theoretical overviews that had been developed in the natural sciences. Nevertheless, the late twentieth century brought striking advances in data collection and conceptualization in the fields of anthropology, sociology, linguistics, economics, psychology, history, and philosophy. While most of their research focused on European-based societies of the past two hundred years, some scholars in each field found ways to expand their exploration of earlier times and to consider larger proportions of the world. Perhaps more difficult were the efforts to open cross-disciplinary conversations within the social sciences and to draw on the astonishing amount of new data produced in natural-science research. Yet it began to be feasible to reopen, on a more informed basis, the unsuccessful debates of the nineteenth century on the relations between biological evolution and human social evolution.

While philosophical flexibility, precision, and discourse arguably advanced in every field of study, ideological shifts in the public arena affected scholarly research. A sharp ideological break—at about 1980—brought neoliberalism, anti-regulation, individualism, and social hierarchy to the fore in public affairs. This change made clear that the preceding era, from about 1945, had sustained quite different outlooks: reaction to world war had generated egalitarianism, a great setback for empires, and massive decolonization—with group consciousness appearing in international organizations, trade unionism, social welfare programs, and socialist movements. Yet the neoliberal era met its complexities too: globalization brought the collapse of the Soviet Union and the end of the Cold War but also the greatest-ever upheaval of popular consciousness in the democratization movements of 1989–1992.

The economics profession, which gained great influence in academia and in politics of the great powers during the era of macroeconomic policy, shifted its leading outlook to a focus on microeconomic analysis and sharp criticism of economic regulation. The views of Friedrich Hayek and Milton Friedman gained wide recognition; they contributed to the rise of a corporate ideology that eventually came to be called "neoliberalism." Game theory approaches gained increasing application, focusing on individual analysis in an age of massive changes at the group level. The rise of Margaret Thatcher and Ronald Reagan to national leadership in the United Kingdom and United States, in 1980 and 1981, brought governmental support to this vision of economics and policy. Issues in environment and discrimination were to be resolved by the price mechanism rather than legislative policy; regulations or limitations on corporate practices were to be reduced or eliminated.

Cross-Disciplinary Scales, Tools, and Debates. As the experience with multiple scales spread from physical to biological sciences and then to social sciences, a general understanding of the issue and the terminology of "scale" began to develop. Population thinking, addressing the fine-structure of units at each scale, became seen increasingly as an aspect of the overall concept of scale. Biological analysis diversified with the molecular work in genomics, the larger scale of epigenetics, and the social scale of public health. At the border of biological and social science studies, the new conception of cultural evolution expanded, with research focused on social learning at an individual level and seeking to rely on population thinking. In social sciences, accelerated collection of data encouraged comparison and linkage of many societies across space, time, and scale; disciplines from economics to art history experimented with exchanging data and concepts. The rise to prominence of rational choice theory, in practice, raised walls between social science and biological analysis. The emerging field of world history, with growing breadth of interest in exploring a comprehensive study of the human past, inched into contact with the physical and biological sciences.[1]

Game theory and systems analysis, modeled at small and large scales respectively, gained application in widely ranging fields of study. Game theory models became influential in several biological subfields, in economics, and in cultural evolution; rational choice theory amplified individual-level thinking with slightly different assumptions. Systems thinking was first influential in fields of engineering and economics; even there its initial efforts were rather shallow in terms of scale. Yet systems thinking persisted in local versions within multiple disciplines and gained new interest as the Santa Fe Institute opened in 1984. This multidisciplinary research institute began systems research and later established an annual Complex Systems Summer School, where "complex systems" were seen to exhibit hierarchical self-organization under selective pressures. Among the dynamics that gained early attention in complex systems was chaos theory, addressing behavior of dynamical systems that are highly sensitive to initial conditions.[2]

Theories of "rational choice" gained wide attention from the 1980s in the fields of sociology, economics, and political science. This approach was

---

[1] Mary C. Stiner, Timothy Earle, Daniel Lord Smail, and Andrew Shyrock, "Scale," in Shryock and Smail, eds., *Deep History*, 242–72.

[2] L. Douglas Kiel and Euel Elliott, eds., *Chaos Theory in the Social Sciences: Foundations and Applications* (Ann Arbor: University of Michigan Press, 1997).

at once a reaffirmation of individualistic analysis and a fundamental critique of individual-based social science. In launching the approach, sociologist James S. Coleman gave a forceful critique of Talcott Parsons' effort to develop a general theory of social action. Coleman argued that Parsons and subsequent scholars had found no way to move from individual-level behavior up to aggregate levels. As a response, Coleman proposed the replacement of economist-style psychological assumptions for the behavior of individuals with general utility functions that could be specified as appropriate for any discipline. Then, Coleman proposed multiple avenues of analysis reaching intermediate levels of aggregation.[3] In principle, rational choice analysis gave attention to intermediate-level structures (such as institutions) through which individuals worked to mediate their place in society. This was formally an argument for attention to multiple scales and variance in social situations. Economist Gary Becker joined with James Coleman, theorizing mid-level behavior by proposing analyses of discrimination against social minorities, of crime, of formation of human capital, and economic analysis of family behavior. This approach came to be known as "rational choice theory," and it met with enthusiastic adoption in economics, sociology, and political science.[4] In practice, a wave of new research did not bring a rapid consensus either on the nature of individual utility functions or on intermediate-level connections between individuals and society as a whole. Rational choice analysis, in practice, was ironically absorbed into the expanding paradigm of neoliberal thinking, with its emphasis on individualistic philosophy and principled opposition to regulation of private institutions.

Game theory reached a wide audience through the work of Robert Axelrod, with his argument that individualism could advance cooperative behavior. Axelrod successfully popularized the reasoning enunciated by W. D. Hamilton and John Maynard Smith in a 1984 book, *The Evolution of Cooperation*.[5] Axelrod carried out large-scale experiments with the Prisoner's Dilemma game and concluded that the Tit for Tat strategy was the most effective: this strategy defects on the first move, and thereafter makes the same move as did the opponent on the previous move. Axelrod showed that this strategy could lead over time to expansion of implicit cooperation, and hence to "evolution." This insightful popular book arguably anticipated later studies of "ultrasociality," exploring large-scale cooperation and reciprocity in the framework of cultural evolution. Axelrod's book, engaging and popular in business communities, perhaps came precisely at the right moment to reinforce expanding neoliberalism.

---

[3] Udehn, *Methodological Individualism*, loc. 6720; Coleman, "Social Theory, Social Research, and a Theory of Action," *American Journal of Sociology* 91 (1986): 1309–35.

[4] Gary Becker, "The Economic Way of Looking at Life," https://www.nobelprize.org/uploads/2018/06/becker-lecture.pdf.

[5] Robert Axelrod, *The Evolution of Cooperation* (New York: Basic Books, 1984).

In contrast to these examples of innovative work on individual behavior and its ironic reception, two philosophers led in innovating thinking about group behavior. In the 1990s, philosophers John Searle and Raimo Tuomela began a long interchange on formulations of "collective intentionality," in which they defined an "institution" as a group of people with a common objective and shared norms. Searle's 1995 book applied the notion of collective intentionality to the study of institutions and argued that syntactic language is the most basic of institutions; he extended this thinking to a general theory of institutions and institutional facts.[6] This approach distinguishes an "I-mode" and a "we-mode" of social thinking and behavior, institutionalized as distinctive "I-groups" and "we-groups."[7] In I-mode, participants in a group focus on their own interests; in we-mode, participants in a group share a group objective. Tuomela later elaborated this analysis more fully.[8] Earlier work by Peter M. Blau can be seen as prefiguring this analysis: Blau distinguished microsociological from macrosociological analysis and envisioned exploring the interaction of the two.[9]

Physical sciences. The field of climatology had achieved considerable sophistication by 1990, and the results could now be conveniently summarized. William F. Ruddiman, in a well-known textbook, classified the eras and arenas of climatological study into the early days of the faint young sun, the long-term study of climate through plate tectonics, the more recent study of climate through orbital-scale changes, and the studies of ice cores to analyze climate change since the last glacial maximum.[10] For earlier times, geologists had learned of the addition of water to the Earth in the early days of the planet and of the era from 715 to 595 million years ago, which may have been a time of a very cold "snowball Earth." The archives of evidence on

[6] John Searle, *The Construction of Social Reality* (New York: The Free Press, 1995), 59–126; Raimo Tuomela, *The Philosophy of Social Practices: A Collective Acceptance View* (Cambridge: Cambridge University Press, 2002), 156–200; Searle, *Making the Social World: The Structure of Human Civilization* (New York: Oxford University Press, 2010), 90–122.

[7] Raimo Tuomela, *Social Ontology: Collective Intentionality and Group Agents* (New York: Oxford University Press, 2013); David P. Schweikard and Hans Bernhard Schmid, "Collective Intentionality," *The Stanford Encyclopedia of Philosophy* (Summer 2013 Edition), Edward N. Zalta, ed., https://plato.stanford.edu/archives/sum2013/entries/collective-intentionality/.

[8] Tuomela relied especially on Bacharach; it may be that Bacharach's initiative in game theory of teams drew on Coleman's proposals. Tuomela, *Social Ontology*; Bacharach, *Beyond Individual Choice*; Coleman, "Social Theory."

[9] Blau, seeking to model both microsocial and macrosocial behavior in human interaction, found the need to define different exogenous variables in each case, and therefore to construct separate theories for individual and group behavior. Thus, Blau's work on "exchange" among individuals can be considered as an analysis of the patterns of I-groups, while his view of macrosocial behavior can be a focus on the patterns of we-groups. Peter M. Blau, "Introduction to the Transaction Edition," *Exchange and Power in Social Life* (New Brunswick, NJ: Transaction Publishers, 1986), vii–xvii.

[10] William F. Ruddiman, *Earth's Climate, Past and Future*, 2nd ed. (New York: W. H. Freeman, 2008).

historical climate included sediments in oceans, existing and previous lakes, glacial ice, and in the rings of trees—reconfirming the close ties of biotic evidence (remains of plankton and pollen) and geochemical evidence.

International conferences considered the reports of the IPCC and proposed ways to limit the steadily increasing atmospheric temperature. But some governments and most powerful corporate interests claimed to be skeptical about the scientific results, and successfully insisted on delays in any policies to limit environmental degradation. Thus arose another major struggle—a social struggle over implementing policy, in the aftermath of the analytical struggle to learn about the processes of climatic change.[11]

The debates on the Gaia thesis, after three decades of discussion, took two new turns as the next century opened. First, James Lovelock himself changed his position on the sustainability of Gaia. He had initially emphasized the stability of Gaia—that its planetary system of organic–inorganic interactions kept average temperate within a range of 10 °C, with parallel limits on humidity. As a result, life forms changed substantially over time but the system underlying their lives remained robust. Lovelock had thought that human society, no matter how serious its interventions, should be considered a part of Gaia and its transformations rather than as a threat to Gaia. But in a 2007 book, he expressed concern about the expanding threats to the atmosphere and to plant and animal species.[12] The second turn in debate was that the field of geology, having continued to expand its knowledge in large-scale and long-term analysis of the Earth system, found itself in general agreement with the Gaia hypothesis, and incorporated its thinking into Earth sciences.[13]

Biology. For the social implications of biological science, the most striking post-1980 research was published in a 1987 article by Rebecca L. Cann,

[11] The Intergovernmental Panel on Climate Change (IPCC, https://www.ipcc.ch/reports/), constituted in 1988 by the United Nations, makes regular reports in climate conditions and prospects. The objections to policy intervention in climate change came not from the scientific community but from economic and political interests outside of climatic analysts. With time and further intervention by these special interests, there gradually developed a small number of global warming skeptics among the scientists. After an initial international environmental conference in Stockholm in 1972, conferences took place in Rio de Janeiro (1992), Kyoto (1997), and Paris (2015).

[12] Lovelock has raised the latter possibility but has treated it as the danger that humanity might have to take on primary responsibility not only for its own welfare but also the larger task of shepherding all planetary processes. James Lovelock, *The Revenge of Gaia: Earth's Climate Crisis and the Fate of Humanity* (New York: Basic Books, 2007).

[13] The 1987 CLAW hypothesis, advanced by four authors (including Lovelock) whose names formed the acronym CLAW, advanced a specific hypothesis on a mechanism linking living plankton to climate, and thereby indicated a growing commonality of the Gaia hypothesis and Earth sciences. Robert J. Charlson, James E. Lovelock, Meinraj O. Andraea, and Stephen G. Warren, "Oceanic Phytoplankton, Atmospheric Sulphur, Cloud Albedo and Climate," *Nature* 326 (1987): 655–61. Nonetheless, critique of the Gaia hypothesis continued, for instance in Toby Tyrrell, *On Gaia: A Critical Investigation of the Relationship between Life and Earth* (Princeton: Princeton University Press, 2013), 208.

Mark Stoneking and Allan C. Wilson, which displayed the results of a remarkable genetic experiment. They analyzed the full extent of the mitochondrial DNA genome for 147 individuals selected to be representative of the overall human population. These results indicated that humanity had formed in Africa and spread to other regions.[14] More precisely, the results indicated common ancestry for all the subjects and that their earliest common female ancestor lived some 200,000 years ago in Africa, where the greatest genetic diversity was concentrated. Results of this research were widely understood to confirm the unity of the human population and opened up interest in long-term history.[15] The results also affirmed that the perceived racial distinctions of present-day humanity were of little significance, thereby encouraging a search for a common mechanism of social development. Genomic analysis then accelerated, focusing on the initially feasible small-scale analyses of mitochondrial DNA (tracing the female line) and Y-chromosomes (tracing the male line), providing results that led to rapid development of interpretations of early human migrations. At a larger scale, plans began immediately for efforts to sequence the full human genome. The initial project, supported by the US government, began in 1990 and announced its results in 2003.[16]

Parallel work in epigenetics rapidly broadened and restructured biological studies of evolution. The evo-devo breakthrough opened debates at the boundary of population genetics and epigenetics. Three overlapping debates arose during the 1970s and have yet to be resolved. The debates focus, respectively, on "inclusive fitness" as a standard for evaluating biological evolution, "reductionism" as a strategy and ideology of scientific analysis, and "human nature" as an encompassing characterization of behavior. In the first of these, the notion of "inclusive fitness" asserts that the standard for successful propagation of genetic inheritance gives attention to the number of offspring not only of individuals but of their close relatives. This standard became significant in studies of cultural evolution, especially because it opened the door to interaction of biological and social processes.[17] Edward O. Wilson, a specialist in the social behavior of ants, argued in contrast that epigenetics, beyond providing a distinctive process of ontogenic development, opened a path by which genetic ancestry exercised direct control over social behavior. He proposed a synthesis of "sociobiology"—followed up by

[14] The 147 individuals were almost all residents of California, where the research was done, but replication of the results with larger and more diverse groups of subjects showed the results to be robust. Rebecca L. Cann, Mark Stoneking, and Allan C. Wilson, "Mitochondrial DNA and Human Evolution," *Nature* 325 (1987): 31–36.

[15] For an especially insightful overview of human biodiversity showing the importance of this research, see Jonathan Marks, *Human Biodiversity: Genes, Race, and History* (New York: Walter de Gruyter, 1995), 169–72.

[16] US government project by National Institutes of Health. A parallel private program sponsored by Celera Genomics distributed its work among twenty leading global universities.

[17] Hamilton, "Genetical Evolution"; John Maynard Smith, "Group Selection and Kin Selection," *Nature* 201 (1964): 1145–47.

a theory of "gene-cultural coevolution"—in which the pressures of natural selection brought not only adaptation of animals to the natural environment but also the genetic evolution of advantageous social behavior. As Wilson saw it, a focus on inclusive fitness and the sharing of genes among siblings added nothing of value to the tracing of ancestry from parent to direct offspring.[18] This major split among researchers continued into the twenty-first century: in a 2010 article in *Nature*, Martin Nowak and E. O. Wilson rejected the validity of W. D. Hamilton's notion of "inclusive fitness" and, with it, the notions of kin selection and altruistic behavior. Arguing that "standard selection theory" represents "a simpler and superior approach," they thus rejected the premise of studies in the expanding field of cultural evolution.[19] A brief but forceful response led by Patrick Abbott and over a hundred co-authors reaffirmed the value of inclusive fitness theory.[20] Further publications continued on each side, in a debate that, basically, put Nowak and E. O. Wilson (along with Richard Dawkins) on the side of relying entirely on direct genetic inheritance, and most others on the side of inclusive fitness, which allowed some role for social processes in evolution.[21]

In the second debate, the earlier debates on "reductionism" took new form, as Wilson's vision of sociobiology met opposition from Stephen Jay Gould and from a group led by Richard Lewontin, each challenging Wilson's view of sociobiology as reductionist and oversimplified. In their 1984 book, Lewontin, Steven Rose, and Leon Kamin—two biologists and a psychologist—challenged Wilson's sociobiology project, arguing that it was ideologically aimed at reducing social issues and social behavior to biological elements.[22] Thus, they expressed a clear belief that social evolution existed and that biological

---

[18] Edward O. Wilson, *Sociobiology: The New Synthesis* (Cambridge, MA: Belknap Press of Harvard University Press, 1975); Edward O. Wilson, *On Human Nature* (Cambridge, MA: Harvard University Press, 1978); Charles J. Lumsden and Edward O. Wilson, *Genes, Mind, and Culture: The Coevolutionary Process* (Cambridge, MA: Harvard University Press, 1981).

[19] Martin A. Nowak, Corina E. Tarnita, and Edward O. Wilson, "The Evolution of Eusociality," *Nature* 466 (2010): 1057–62; Hamilton, "Genetical Evolution."

[20] Patrick Abbott, et al., "Inclusive Fitness Theory and Eusociality," *Nature* 471, 7339 (2011): E1–4. For a more detailed response, see Regis Ferriere and Richard E. Michod, "Inclusive Fitness in Evolution," *Nature* 477 (2011): E6–E8.

[21] A. Gardner, S.A. West, and G. Wild, "The Genetical Theory of Kin Selection," *Journal of Evolutionary Biology* 24 (2011) (5): 1020–43. For an expanded restatement of the critique of inclusive fitness, see Benjamin Allen, Martin A. Nowak, and Edward O. Wilson, "Limitations of Inclusive Fitness," *PNAS* 110, 50 (2013): 20135–39.

[22] Gould, *Ontogeny and Phylogeny*; Lewontin, et al., *Not in Our Genes*. Neither Gould nor Lewontin et al., however, proposed details on principles of social evolution. Lewontin et al. argued that the ideological implication of sociobiology was that social activism of any sort was futile, since all of life is governed at the biological level. For an interview with Lewontin that pursues this issue, see R. C. Lewontin, Diane Paul, John Beatty, and Costas B. Krimbas, "Interview of R. C. Lewontin," in Rama S. Singh, Costas B. Krimbas, Diane B. Paul, and John Beatty, eds., *Thinking About Evolution: Historical, Philosophical and Political Perspectives*, Vol. 2 (Cambridge: Cambridge University Press, 2001), 22–61.

evolution should not be proposed to substitute for it, although they did not identify specific processes of social evolution. The debate led to a 1997 symposium on biological reductionism in London, at which Wilson was not represented.[23] Michel Morange, in his skillful analysis of research in the rise of molecular biology, gave ample attention to debates on reductionism and concluded in general that strict reductionism could not be upheld, so that multiple levels of analysis will continue to be necessary: "The recourse to higher levels of analysis is utterly indispensable if the edifice of modern biology is to remain intact."[24] Morange thus implied that biology, to preserve its disciplinary integrity, would need to make space for a field of social evolution. Biologists took no active steps in that direction and social scientists did little more.

The third debate brought new discussion on another old issue: "human nature." The initial salvo was by E. O. Wilson, whose *Human Nature* argued that genetic and epigenetic change would be able to explain every dimension of human behavior—thus, human nature is located entirely at the individual and biological level. Richard Lewontin et al. argued in contrast that social behavior is determined by social rather than biological factors—so that human nature is located primarily at the social level, as it was seen by Boas. Sociologist Joseph Lopreato sought to bridge the gap with a 1984 book arguing that human nature is based at both individual and group levels.[25] Steven Pinker later devoted a book to rejecting the notion that human behavior depends entirely on nurture rather than nature, affirming that biology is essential to human nature. He traced this notion to the "blank slate" of John Locke, arguing that twentieth-century social sciences in general had mistakenly adopted Locke's assumption. Lisa Feldman Barrett returned to the issue in her analysis of emotion, emphasizing that emotions are constructed in interactions among individual humans rather than inherited biologically.[26] These assertions have been passionate but inconclusive, in that none have been able to define human nature specifically nor to establish whether it is based at biological or social, individual or group levels. Yet it seems that debates on human nature will continue.

Biology's great theoretical advances of the late twentieth century were in genetic analysis (details of the genome itself and of gene-protein links), in epigenetic analysis (ways in which proteins modify genetic processes), and the combination of genetics and ontogenic development through evo-devo

---

[23] Novartis Foundation Symposium, *The Limits of Reductionism in Biology* (New York: John Wiley & Sons, 1998).

[24] Morange, *History of Molecular Biology*, 246.

[25] Wilson, *Human Nature*; Lewontin, et al., *Not in Our Genes*; Joseph Lopreato, *Human Nature and Biocultural Evolution* (Boston: Allen & Unwin, 1984). Lopreato's work includes significant attention to mechanisms of social evolution.

[26] Steven Pinker, *The Blank Slate: The Modern Denial of Human* Nature (New York: Viking, 2002); Barrett, *How Emotions Are Made: The Secret Life of the Brain* (Boston: Houghton Mifflin Harcourt, 2018).

thinking.[27] This expansion of biological theory provided openings for incorporating innovations from other disciplines into the understanding of biological evolution. Thus, through ingenious chemistry, genomes could now be sequenced for humans and many other species with rapidity and precision. New knowledge from geology brought discovery in biology, as study of plate tectonics and geological strata showed several past waves of extinction among species of all orders and times—so that the term "sixth extinction" was coined to emphasize that extinctions during the current Anthropocene era were shaping up to be of a magnitude parallel to those of the preceding waves.[28] Advances in historical genetics built pressure for paleontological research on tropical regions that yielded important new data, especially for Africa.[29] Game theory was central to the conceptualization of inclusive fitness and eventually became central to conceptualizing group behavior. Advances in cladistic analysis—graphical display of typologies—helped to identify past linkages and separations among genetic groups. Various versions of a "molecular clock" were developed to date the divergence of genetic groups.[30] Further topics that could now be explored at higher levels of biological sophistication included social learning, linguistics, emotions, and gesture—these were taken up by scholars based in the social sciences.

Two new discourses developed on the size of populations: a genetic analysis of population sizes and an anthropological analysis of human and primate group sizes. First, geneticists developed a method for estimating current and past population sizes. The key variable is the "effective population size" N_e, the number of breeding individuals who contribute to the population of the next generation. Working even with a single individual genome, N_e may be estimated based on linkage disequilibrium between genetic markers; other techniques rely on larger amounts of data and different algorithms. Several estimates have been made of human populations in various times and places.[31] While the technique will surely be valuable, a recent review

---

[27] Gould, *Ontogeny and Phylogeny*; Jacob, "Evolution and Tinkering."

[28] Paul B. Wignall, *The Worst of Times: How Life on Earth Survived Eighty Million Years of Extinctions* (Princeton: Princeton University Press, 2015). Richard Leakey and Roger Lewin, *The Sixth Extinction: Patterns of Life and the Future of Humankind* (New York: Doubleday, 1995); Elizabeth Kolbert, *The Sixth Extinction: An Unnatural History* (New York: Henry Holt, 2014).

[29] Chris Stringer and Robin McKie, *African Exodus* (London: Cape, 1996); Chris Stringer and Peter Andrews, *The Complete World of Human Evolution* (London: Thames and Hudson, 2005).

[30] The design of a "molecular clock" required making social, historical, and genetic judgments. DNA samples were taken from living humans, but the analysts then had to link these living individuals to presumed population groups in the past, by making assumptions about their racial, ethnic, linguistic, or migratory ancestry. Then rates of genetic change over time had to be estimated. Techniques are still in need of significant advance.

[31] S. H. Ambrose, "Late Pleistocene Human Population Bottlenecks, Volcanic Winter, and Differentiation of Modern Humans," *Journal of Human Evolution* 34, 6 (1998): 623–51; Jinliang Wang, "Estimation of Effective Population Sizes from Data on Genetic Markers," *Philosophical Transactions of the Royal Society of London B Biological Sciences* 360, 1459 (2005): 1395–409.

confirms that there remain many questions on how to interpret the estimates generated in this analysis.³² Second, anthropologists Leslie Aiello and Robin Dunbar documented the distribution of average group sizes of contemporary primate species, compared them to average neo-cortex capacity, and found a strong correlation. They found a parallel correlation in cross-sectional correlation of grooming time and neo-cortex size for existing primates. They used these relationships to project the group size for early hominin species, based especially on neo-cortex size. They concluded that the average group size for *Homo sapiens* should be 150—and found that groups of 150 persons are very common among modern humans.³³ Dunbar, in further studies, noted varying types and sizes of human groups, including bands and language groups.³⁴ In sum, both genetic and anthropological estimates of evolving group sizes are advancing, but the issues surrounding both methods remain complex, and we have little empirical data with which to check the estimates.

The end of the twentieth century provided an appropriate moment for two distinguished biologists to show that, despite debates from inclusive fitness to group size, evolutionary thinking had developed impressively. John Maynard Smith and Eörs Szathmáry published a work synthesizing contemporary evolutionary thinking, working from the beginnings of eukaryotic cells to human evolution.³⁵ This concise and clearly written volume traced the transitions from chemistry to life and to each new scale and type of life, showing how knowledge about each level has helped explain overall patterns in life.

Cultural Evolution and mild reductionism. Two pairs of researchers pursued their earlier efforts to define mechanisms by which social processes might be liked to genetic evolution. Cavalli-Sforza and Feldman, followed soon by Richerson and Boyd, took inspiration from Darwinian theory, population genetics, and social learning theory in work to find a new path to human development. They adopted forms of reductionism that were milder than the sociobiology of E. O. Wilson—who argued that the Darwinian mechanism of natural selection was sufficient to explain human social structures as well as physical phenotype. The two groups worked on parallel tracks for roughly twenty years. In effect, these proponents of cultural evolution argued that they had identified a previously unknown facet of evolution. In it, individual-level human choices and learning activities, retained in the brain through processes that have some analogy to biological evolution, also

---

[32] M. Husemann, F. E. Zachos, R. J. Paxton, and J. C. Habel, "Effective Population Size in Ecology and Evolution," *Heredity* 117 (2016): 191–92; see also the associated special issue.

[33] Leslie C. Aiello and R. I. M. Dunbar, "Neocortex Size, Group Size, and the Evolution of Language," *Current Anthropology* 34 (1993): 184–93.

[34] Robin Dunbar, *Grooming, Gossip and the Evolution of Language* (London: Faber and Faber, 1996); Robin I. M. Dunbar and Richard Sosis, "Optimising Human Community Sizes," *Evolution and Human Behavior* 39 (2018): 106–11.

[35] John Maynard Smith and Eörs Szathmáry, *The Major Transitions in Evolution* (Oxford: Oxford University Press, 1995).

convey a population dimension so that they contribute to social behavior.[36] The analysis argued that biologically based individual-level dynamics of cultural evolution can explain types of behavior that have previously been argued to be results of group behavior. I call it "reductionist" because it models culture in terms that are fitted as closely as possible with biological change, and it extends the resulting models as far as possible into cultural affairs. Initial work began with mathematical modeling of individual choices in learning, relying on a game-theoretical or population-genetics approach to expand the notion of "social learning," first developed in psychology.[37] For some time, the models were abstract and general, but they gradually developed hypotheses on the role of social learning among hominin of the Pleistocene era.

How did these scholars define "culture"? How did the term "cultural evolution" arise? In their initial 1976 publication, Boyd and Richerson defined "culture" as follows: "The analysis which follows addresses only a minimal, reduced conception of culture: the information learned by individuals which affects or potentially affects phenotypic behavior and which is acquired from other individuals. It is only this narrowly defined, although very fundamental aspect of culture, that is comparable to the genetic system of individuals."[38] Cavalli-Sforza and Feldman used the term "cultural inheritance" in their earliest work (1973), then highlighted the term "cultural evolution" beginning in 1978.[39] In the first book-length theoretical statement in cultural evolution (1981), Cavalli-Sforza and Feldman defined their framework, saying "We will use the term 'cultural' to apply to characteristics that are learned by any process of nongenetic transmission, whether by imprinting, conditioning, observation, imitation, or as a result of direct teaching. ... A culturally acquired behavior becomes part of the overall *phenotype*."[40] Their research also focused on rates of diffusion of innovations and on theories of transmission of cultural characteristics, distinguishing vertical transmission (parent-to-child)

---

[36] "[B]oth genes and culture are informational entities that are differentially transmitted from one generation to the next .... [T]he population genetics approach regards culture as an evolving pool of ideas, beliefs, values, and knowledge that is learned and socially transmitted between individuals." Kevin N. Laland, "Cultural Evolution," in Mark Pagel, ed., *Encyclopedia of Evolution*, Vol. 1 (Oxford: Oxford University Press, 2002), 220.

[37] Bandura, *Social Learning Theory*.

[38] Robert Boyd and Peter J. Richerson, "A Simple Dual Inheritance Model of the Conflict Between Social and Biological Evolution," *Zygon, Journal of Religion and Science* 11 (1976): 129.

[39] L. L. Cavalli-Sforza and M. W. Feldman, "Towards a Theory of Cultural Evolution," *Interdisciplinary Science Reviews* 3 (1978): 99–107. Cavalli-Sforza was certainly widely read in the social science literature and aware of the various uses of the terms "cultural" and "social" evolution. I continue to seek out citations to see whether earlier publications used the term "cultural evolution." Note that anthropologists White, Service, and Trigger used the term "social evolution," but to refer to group rather than individual behavior. E. O. Wilson, however, used the term "cultural evolution" as an element of his genetically driven sociobiology project.

[40] Cavalli-Sforza and Feldman, *Cultural Transmission and Evolution*, 7–15.

from partly or fully horizontal transmission, where characteristics may be discrete or continuous. They continued to speak of cultural evolution and, increasingly, "cultural transmission" until their concluding publication on this topic in 1994.[41] Cavalli-Sforza gradually shifted his research from cultural evolution to language, apparently envisioning language as providing an additional linkage of genetic and cultural domains. In a widely cited 1988 article on "The Reconstruction of Human Evolution," he emphasized parallels of genetic, linguistic, and archaeological data in confirming major divisions within human populations; he relied on Greenberg's classifications for language data.[42] In a 1990 response and debate, both anthropologists and some biologist commentators expressed skepticism at the global scale of the analysis.[43] Cavalli-Sforza went on to publish several more works developing this theme: his initiative was pursued by biologists but without active collaboration of anthropologists, linguists, or cultural evolution scholars.

Boyd and Richerson, in their 1985 book, described their focus as, "Darwinian theory of the evolution of cultural organisms," which they called "dual inheritance theory" because it traced both cultural and genetic inheritance. As they argued, "Culturally acquired variations are transmitted from generation to generation and, like genes, they are also evolving properties of the population. ... Because of this we will say that culture has 'population-level consequences.'" They emphasized that social learning causes the communication of phenotypic characteristics directly from individual to individual; they traced patterns of transmission that they called "guided variation" and "biased transmission."[44] They argued that cultural evolution and genetic processes are different processes, operating by different mechanisms, for which the processes of selection may not give the same results: in that sense, they are in competition.[45] Analysis therefore explores whether there can be an equilibrium in which the characteristics resulting from cultural and genetic evolution are consistent with one another. They argued that

---

[41] M. W. Feldman, L. L. Cavalli-Sforza, and L.A. Zhivotovsky. "On the Complexity of Cultural Transmission and Evolution," in G. Cowan, D. Pines, and D. Meltzer (eds.), *Complexity: Metaphors, Models, and Reality* (Boston: Addison Wesley, 1994), 47–62. But see also their later encyclopedia article: M. W. Feldman and L. L. Cavalli-Sforza, "Cultural Transmission," in M. Pagel, ed., *Encyclopedia of Evolution* (New York: Oxford University Press, 2002), 222–26.

[42] Cavalli-Sforza, L. L., Alberto Piazza, Paolo Menozzi, and Joanna Mountain. "Reconstruction of Human Evolution: Bringing Together Genetic, Archaeological, and Linguistic Data," *PNAS* 85, 16 (1988): 6002–6006.

[43] Richard Bateman, et al., "Speaking of Forked Tongues: The Feasibility of Reconciling Human Phylogeny and the History of Language [and Comments]," *Current Anthropology* 31, 1 (1990): 1–24. In my view, Cavalli-Sfroza's path-breaking energy and imagination were important in advancing study of human evolution, yet there were limits to the specifics of his arguments on blood types, migration patterns, cultural evolution, and links of language and ethnicity.

[44] Boyd and Richerson, *Culture and the Evolutionary Process*, 2–6.

[45] Boyd and Richerson, *Culture and the Evolutionary Process*, 172–203. The subtitle of this chapter is "conflicts between cultural and genetic evolution."

their theory, while focused on individual behavior, is of relevance beyond that level.[46] Boyd and Richerson continued discourse with Campbell; their 1985 work includes a section on "group selection and the evolution of cooperation." Campbell, in his last intervention on this topic, backed away somewhat from his earlier emphasis on group-based social evolution, acknowledging the work of Hamilton and D. S. Wilson on kin selection and structured demes. But he reaffirmed his position by defining "ultrasociality" as "that high level of sociality in which full-time division of labor occurs, with specialized roles whose occupants do no food gathering and are fed by others"—the social insects and urban humankind. Campbell argued, though with little detail, that practices envisioned in the Boyd and Richerson array of activities provide a "route to ultrasociality."[47] This term was to have echoes in later years, though its meaning was to be transformed.

Boyd and Richerson first used the term "cultural evolution" in a 1996 title: "Why Culture is Common but Cultural Evolution is Rare." From 2000, the term "cultural evolution" became common in their usage, and it was adopted by their colleagues as their number increased. By that time, Boyd and Richerson had outlined the full framework of their theory of cultural evolution. The dual-heritage model of social learning traced decision rules in which learners chose between learning behavior from environmental cues or copying behavior from social cues of model individuals. If environmental conditions of one sort or another were stable enough, one or the other strategy proved to be best, so that natural selection would lead that strategy to expand within a population.[48] As a result, in some cases, long-term application of "conformist transmission rules" brought growth of populations that

---

[46]"In our theory, the individual is linked to larger units by cultural transmission and its population-level properties. What this linkage implies for what behavior we can expect (e.g. should the function of behavior be interpreted at the individual or group or group level) turns out to depend on the details of evolutionary processes." Boyd and Richerson, *Culture and the Evolutionary Process*, 17.

[47]Campbell, "The Two Distinct Routes Beyond Kin Selection to Ultrasociality: Implications for the Humanities and Social Sciences," in Diane L. Bridgeman, ed., *The Nature of Prosocial Development: Interdisciplinary Theories and Strategies* (New York: Academic Press, 1983), 37.

[48]Boyd and Richerson, "Why Culture Is Common But Cultural Evolution Is Rare," *Proceedings of the British Academy* 88 (1996):73–93. They used the term "social evolution" only in their initial 1976 publication. Robert Boyd and Peter J. Richerson, *The Origin and Evolution of Cultures* (New York: Oxford University Press, 2005); Stephen Shennan, *Genes, Memes and Human History: Darwinian Archaeology on Cultural Evolution* (New York: Thames and Hudson, 2002), 35, 84; Peter J. Richerson and Morten Christiansen, eds., *Cultural Evolution: Society, Technology, Language, and Religion* (Cambridge, MA: MIT Press, 2013), 3–11. Perhaps ironically, researchers on such non-human populations as birds began adopting the analysis of Boyd and Richerson, but identified their work as "social evolution," presumably because they were reluctant to apply the term "culture" to their subject. Tamás Székely, Allen J. Moore, and Jan Komdeur, eds., *Social Behavior: Gene, Ecology and Evolution* (Cambridge: Cambridge University Press, 2010).

were relatively strong in learning skills and cooperative behavior; they tended to expand at the expense of other populations.

When did cultural evolution begin? For Pleistocene times, Boyd and Richerson assumed that, from about 250,000 years ago, the population effects of social learning had reached the point where cumulative cultural evolution began.[49] From this point, "cultural group selection" took hold. Cooperative outlook led to a "tribal instinct," encouraging families to form expanded groups—"tribes" or "institutions"—although consolidation of these groups required many thousands of years of genetic and cultural coevolution.[50] For cultural group selection to advance, groups needed to be relatively homogeneous within, but with significant heritable variation among competing groups. It was the competition and selection among groups that led to greater species-level fitness and cooperation. For the Holocene era, starting some 10,000 years ago, Boyd and Richerson assumed that the cumulative effects of cultural evolution had been sufficient to launch creation of larger groups or institutions, along with the rise of agriculture.[51] The groups emerged by combination of tribes through such "work-arounds" as coercive dominance, segmentary hierarchy, and exploitation of symbolic systems. The speed of cultural evolution now accelerated, so that formation of the new institutions could take place in roughly 1000 years, a time-frame parallel to that of the biological evolution of lactose tolerance in dairying population.[52] Thus, from the year 2000 on, cultural evolution was firmly established as a research field; its practitioners became confident in using the term "culture" to refer to individual-level learning, facilitated at times by teaching. As noted earlier, I have chosen to label this aspect of culture as individual-level culture, to distinguish it from group-level culture or cultural exchange at the group level.

---

[49] "We hypothesize that the long-continued effort of cultural group selection in the Pleistocene led to the evolution of the old, tribal, social instincts." Boyd and Richerson 1999, 255–57. For parallel arguments by Joseph Henrich, see J. Henrich and R. McElreath. "The Evolution of Cultural Evolution," *Evolutionary Anthropology* 12 (2003): 123–135; Joseph Henrich, "Cooperation, Punishment, and the Evolution of Human Institutions," *Science* 312 (2006): 60–61; Joseph Henrich, Robert Boyd, and Peter J. Richerson, "Five Misunderstandings About Cultural Evolution," *Human Nature* 19 (2008): 119–37. By 2016, Henrich was ready to push that date back to 750,000 years ago. Henrich, *The Secret of Our Success: How Culture Is Driving Human Evolution, Domesticting our Species, and Making us Smarter* (Princeton: Princeton University Press, 2016), 293.

[50] The tribe is defined as "any institution that organizes interfamilial cooperation." Such groups are argued to have escaped the limits earlier identified by G. Williams, though their dynamics are not otherwise described. Boyd and Richerson, *Origin and Evolution*, 262.

[51] "We believe that the human capacity to live in larger-scale forms of tribal social organization evolved through a coevolutionary ratchet generated by the interaction of genes and culture." Boyd and Richerson, *Origin and Evolution of Cultures*, 263, 265–69.

[52] Peter J. Richerson and Robert Boyd, "The Evolution of Human Ultra-Sociality," in Irenäus Eibl-Eibisfeldt and F. Salter, eds., *Indoctrinability Ideology, and Warfare*, Vol. 13 (New York: Berghahn Books, 1998), 71–95.

Boyd and Richerson further expanded details of their analysis and gave support to empirical analysis. The most comprehensive single statement on cultural evolution was the 2007 volume by McElreath and Boyd, which presented the full evolution of the framework in terms of mathematical models that highlighted the assumptions at each stage.[53] It made clear the argument for multi-level selection, where tribal groups can serve as a basis for evolutionary selection. That is, they addressed group behavior without leaving the logic of individual decisions. These analyses laid groundwork for subsequent work on the Richerson–Boyd vision of "ultrasociality," in which they gradually expanded the scope of application of cultural evolution theory; in a 2010 study, they spoke of "a massive capacity for culture," yet did not go as far as to explore what I have labeled as group-cultural practices or behavior.[54] The term "cultural learning" came to refer to a learning practice more complex than social learning because it required teaching, and which applied to late Pleistocene and Holocene times.[55] Their point was to use innovative analysis of individual-level behavior to explain early-Pleistocene learning and evolution, then to extend the analysis to explain late Pleistocene and Holocene evolution.[56] Stephen Shennan's 2009 collection made the case for "evolutionary anthropology" and made a rare but explicit reference to social evolution.[57]

Paradigms related to cultural evolution. In addition, some quite different research programs have developed which, while closely related to cultural evolution, are different in their definitions and approaches from the "dual inheritance" group. All claim a link to Darwinist thinking, but in practice they give varying emphases to variation, selection, and inheritance. William Durham

---

[53] Richard McElreath and Robert Boyd, *Mathematical Models of Social Evolution: A Guide for the Perplexed* (Chicago: University of Chicago Press, 2007). The term "social evolution" appears in the title but not in the text. It appears inconsistent with other work in cultural evolution, but no explanation is offered.

[54] Peter J. Richerson and Robert Boyd, "The Darwinian Theory of Human Cultural Evolution and Gene–Culture Coevolution," in Michael A. Bell, Douglas J. Futuyma, Walter F. Eanes, and Jeffrey S. Levinton, *Evolution Since Darwin: The First 150 Years* (Sunderland, MA: Sinauer Associates, Inc., 2010), 561–88. The term "ultrasociality" was coined by Donald T. Campbell to mean group-based social evolution rather than cultural evolution.

[55] The term had earlier been used by Michael Tomasello et al. See Michael Tomasello, Ann Cale. Kruger, and Hilary Horn Ratner, "Cultural Learning," *Behavioral and Brain Sciences* 16, 3 (1993): 495–511, https://doi.org/10.1017/s0140525x0003123x. see also Robert Boyd, *A Different Kind of Animal: How Culture Transformed Our Species* (Princeton: Princeton University Press, 2018), 46–47, 175–77.

[56] Boyd and Richerson, *Origin and Evolution*; Peter J. Richerson and Robert Boyd, *Not by Genes Alone: How Culture Transformed Human Evolution* (Chicago: University of Chicago Press, 2005); Boyd and Richerson, "Culture and the Evolution of Human Cooperation," *Philosophical Transactions of the Royal Society* 364 (2009): 3281–88; Richerson and Boyd, eds., *Cultural Evolution: Society, Technology, Language, and Religion* (Cambridge, MA: MIT Press, 2013).

[57] Stephen Shennan, "Introduction," in Stephen Shennan, ed., *Pattern and Process in Cultural Evolution* (Berkeley: University of California Press, 2009), 1–18. See also Shennan, *Genes, Memes and Human History*.

stood out as one applying notions of inclusive fitness, kin selection, and altruism; his extensive statement on coevolution yields a theory aimed at touching many issues at once.[58] On the border separating the camps of sociobiology and cultural evolution was Susan Blackmore, whose analysis extends Richard Dawkins' initial formulation of Memes as cultural quanta which replicate themselves from person to person rather than requiring acts of learning.[59]

Evolutionary psychologist Michael Tomasello drew on the evo-devo turn, emphasizing ontogeny or child development. Tomasello had already published on "cultural learning" in 1993.[60] Starting in 1998, he conducted detailed comparisons of development of young primates: humans, chimpanzees, bonobos. He concluded that perception of the physical world developed in parallel for all three species but that perception of social relations developed to a much higher level in humans. He advanced, as a historical hypothesis, that early Homo began food sharing, which opened the door to cooperation. As he argued, humans of 400,000 years ago began to work closely in pairs (for instance, as mates); by 150,000 years ago the population had risen and required working in groups.[61]

While Tomasello's analysis was in many ways linked to that of Boyd and Richerson—they shared in employing terms such as "ratchet effect," "cumulative cultural evolution," and "cultural learning"—yet it also became clear that they were articulating alternative evolutionary mechanisms. Boyd and Richerson emphasized dual heritage and multi-level selection, yielding genetically supported advance in cooperative outlook: in this model, groups and tribes grew slowly but at growing rates. Tomasello emphasized ontogenic processes of development along with collective intentionality, thus giving more attention to intimate social interaction.

In another sort of coevolution explored at the same time, Jared Diamond, a multidisciplinary scholar based in physiology, published a widely read interpretation of the coevolution of humans, crops, and disease agents.[62] Diamond did not formally espouse a mechanism of social evolution but relied

---

[58] For instance, Durham espoused an "ideational theory" of culture emphasizing that "culture is conveyed socially within or between populations," yet he treated culture in practice in terms of Dawkinsian memes. William H. Durham, *Coevolution: Genes, Culture, and Human Diversity* (Stanford: Stanford University Press, 1991). See also William F. Harms, *Information and Meaning in Evolutionary Processes* (Cambridge: Cambridge University Press, 2004).

[59] Susan Blackmore, *The Meme Machine* (Oxford: Oxford University Press, 1999).

[60] The term "social learning" has gradually expanded in scope, as it has come to be understood that animals as well as humans learn through both imitation and instruction. Tomasello and his colleagues sought to introduce the term "cultural learning" to indicate a form of learning at a higher level, perhaps restricted to humans, in which learning focuses on mutual understanding. Tomasello, et al., "Cultural Learning"; Richerson and Boyd, *Not by Genes Alone*.

[61] Michael Tomasello, "Human Culture in Evolutionary Perspective," in M. Gelfand, ed., *Advances in Culture and Psychology* (Oxford: Oxford University Press, 2011), 5–51.

[62] Jared Diamond, *Guns, Germs, and Steel: The Fates of Human Societies* (New York: Norton, 1997).

on population genetics to argue that Eurasian populations grew in response to levels of nutrition from crops and the immunities that Eurasians developed against disease. This analysis did fit the argument that social evolution and demographic expansion began with agriculture.

Studies of language and linguistics. In linguistics, each of the areas of research that arose in the 1950s expanded further, producing a mix of research advances and new debates. These were: biological capability for reasoning and speech; community maintenance of speech; and language universals. On linguistic universals, Bernard Comrie followed up the four-volume collection of 1978 with a concise review of the new field.[63] As he argued, while it may be that human syntactic language evolved one time only, and that all existing languages are descendants of that original language, the field of linguistic universals depends on other mechanisms relying on the inherent dynamics of language.[64] Joseph Greenberg, after having drawn many others into his study of linguistic universals, continued his individual-level work of language classification worldwide. Much of this work was completed by the time of his death in 2001. Greenberg published a 1970 article classifying the Indo-Pacific languages of New Guinea and surrounding regions; a 1987 book classifying languages of the Americas, showing the distribution of related subgroups throughout the hemisphere; and a two-volume 2000 classification of the languages of Eurasia.[65] This corpus of analysis, reconfirmed and revised by the classification work of Christopher Ehret on three African-based language phyla, permits the comparison of language-distribution data with genetic data in tracing patterns of migration.[66] In further classification work, the Santa Fe Institute

---

[63] Comrie had been trained with Chomsky in generative linguistics but turned to working with Greenberg and language universals: his book reflects a balance and comprehension of the two approaches. Bernard Comrie, *Language Universals and Linguistic Typology* (Chicago: University of Chicago Press, 1981), x–xi, 23–27; Greenberg, et al., *Universals of Human Language*.

[64] This provides a good reason for adopting James Hurford's term of "glossogeny" to refer to historical linguistic change—the distinctiveness of the term reflects the distinctiveness of the process itself. As Fitch suggests, "Glossogeny and Phylogeny Can Interact in Important and Unintuitive Ways." Fitch, *Evolution of Language*, 34; see also 77–93.

[65] Greenberg launched his classification of Indo-Pacific languages in part because of an interest in a possible linkage among Indo-Pacific, Australian, Dravidian, Nilo-Saharan, and Niger-Kordofanian. Greenberg's classification of Eurasiatic languages had some similarities to the Nostratic classification proposed in 1903 by Danish linguist Holger Pederson, a classification that continued to be favored by Russian linguists. Joseph H. Greenberg, "The Indo-Pacific Hypothesis (1971)," in Greenberg (ed. Croft), *Genetic Linguistics*, chapter 12; Croft, "Greenberg," 23; Greenberg, *Language in the Americas* (Stanford: Stanford University Press, 1987); Greenberg, *Indo-European and its Nearest Neighbors: The Eurasiatic Language Family*, 2 vols (Stanford: Stanford University Press, 2000).

[66] Christopher Ehret, *A Historical-Comparative Reconstruction of Nilo-Saharan* (Köln: R. Köppe Verlag, c2001); Ehret, *Reconstructing Proto-Afroasiatic (Proto-Afrasian): Vowels, Tone, Consonants, and Vocabulary* (Berkeley: University of California Press, c1995). On Pleistocene migration of language groups, see Manning, "*Homo sapiens* Populates the Earth"; and Patrick Manning and

set up a project on languages in 1994, including Merritt Ruhlen, a student of Greenberg, and Sergei Starotsin, a supporter of the Nostratic hypothesis.[67] At the same time, Greenberg's post-African classifications revealed major disagreements among historical linguists on classification.[68] Relying on Greenberg's approach to the literature in historical linguistics, historian Patrick Manning proposed in 2006 a language-based interpretation of human occupation of the world.[69]

Studies in evolutionary linguistics broadened greatly with the 1990 publication of *Language and Species*, by Derek Bickerton.[70] This work made the case for "protolanguage," an early lexical system of very basic vocalized representations.[71] Bickerton argued that some remnants of that lexicon remain in language and that today's pidgin languages provide insights into how protolanguage once functioned. Yet he also argued that there was a sudden or "catastrophic" shift from protolanguage to syntactic language. Bickerton's model was taken up by biologist John Maynard Smith, who further linked it to the issue of social evolution in his co-authored grand synthesis of the

---

Aubrey Hillman, "Climate as a Factor in Migration and Social Change, 200,000 to 5000 Years Ago," American Historical Association annual meeting, New Orleans (5 January 2013).

[67] The Nostratic hypothesis posits a phylum including all of the language groups of central Eurasia plus Afroasiatic. For an argument that this collection of language families is best seen not as a specific phylum but as a sizeable portion of the proto-Human language, see Christopher Ehret, "Nostratic—Or Proto-Human?" in Colin Renfrew and Daniel Nettle, eds., *Nostratic: Examining a Linguistic Macrofamily* (Cambridge: McDonald Institute for Archaeological Research, 1998), 93–122. The Santa Fe project continued for a time but appears to have closed down. "Evolution of Human Languages: An International Project on the Linguistic Prehistory of Humanity," *ehl.santafe.edu.* Santa Fe Institute.

[68] The issue became public when Greenberg met furious debate from linguists specializing on the Americas but approbation from scholars in other disciplines, as had been the case with his analysis of Indo-Pacific languages. Joseph H. Greenberg, Christy G. Turner II, Stephen L. Zegura, Lyle Campbell, James A. Fox, W. S. Laughlin, Emöke J. E. Szathmary, Kenneth M. Weiss, Ellen Woolford, "The Settlement of the Americas: A Comparison of the Linguistic, Dental, and Genetic Evidence [and Comments and Reply]," *Current Anthropology* 27 (1986): 477–97.

[69] Manning, "*Homo sapiens* Populates the Earth." This article, relying especially on Greenberg's approach, included documentation of the inconsistency of classification principles used by linguistic specialists on various regions.

[70] Derek Bickerton, *Language and Species* (Chicago: University of Chicago Press, 1981). Also in 1990, Pinker and Bloom made a case that generative syntactic theory was compatible with Darwinian evolution: Steven Pinker and Paul Bloom, "Natural Language and Natural Selection," *Behavioral and Brain Sciences* 13 (1990): 707–27.

[71] I use terms as follows: (1) "language" refers to any and all forms of language—spoken or internal, syntactical or not; (2) "i-language" refers to a logical but unspoken language; (3) "visual communication" refers to communication by gestures; (4) "pre-language" refers to spoken but non-syntactic language; (5) "proto-human" refers to the presumed original, syntactic language; (6) "speech," "syntactic speech," and "e-language" refer to syntactic language. In other usage, Bickerton's "protolanguage" refers to nonsyntactic speech.

great transitions in living systems.[72] A wide-ranging 1998 collective volume includes hypotheses by Robert Berwick and Bickerton for rapid or "catastrophic" emergence of syntax, balanced by arguments of other scholars for gradual emergence of syntax.[73]

Noam Chomsky, whose researches on the hypothesis of Universal Grammar and its complexity had long led the field, proposed in 1995 a simplification of his hypothesis. In *The Minimalist Program*, he argued that the rules for syntax, while inherently complex, could be simplified by certain procedures. Chomsky's distinction between *i-language* and *e-language*—internal thinking processes and the external expression of that thought—began to get wider attention.[74] The idea was that i-language and internal thinking held the Universal Grammar that Chomsky had long been theorizing, while e-language, as soon as it began to be spoken, experienced the great variety of human speaking habits. Berwick proposed, in 1998, a specific formulation of these ideas with his notion of "Merge," a hypothesized genetic change enabling humans to link thoughts hierarchically, adding each new item in thought to a previously constructed collection of lexical items.

Large-scale social change. Studies of macro-level social change, though not generally relying on evolutionary mechanisms, continued with a phenotypical emphasis on the external appearances of societies and states. In contrast to all other evolutionary processes, the logic of social evolution was understood by its proponents to depend on conscious human construction of social and institutional forms. Nevertheless, processes of social evolution surely relied as well on the advanced human capabilities arising from increased brain size, as new capabilities interacted with each other and with environmental factors. Further, one must ask whether social evolution, once launched, relied primarily on the dynamics of human individual or group behavior or whether coevolution with other processes was equally central. The analysts of social evolution, even in the years after 1980—and with the principal exception of Donald Campbell—rarely linked their interpretations directly to the various studies of biological or cultural evolution.

Scholars in anthropology and sociology continued the project of macro-level social evolution, as initiated by Leslie White. Two

---

[72] Maynard Smith and Szathmáry, *Major Transitions in* Evolution, 255–309; see also W. Tecumseh Fitch, *The Evolution of Language* (Cambridge: Cambridge University Press, 2010), 401–32.

[73] "What we need therefore is an account of language evolution, sensitive both to language as 'a sort of contract signed by members of a community' and to language as a hard-wired (individual) competence generated under standard processes of Darwinian natural selection." Michael Studdert-Kennedy, Chris Knight, and James R. Hurford, "Introduction: New Approaches to Language Evolution," in Hurford, Studdert-Kennedy, and Knight, eds., *Approaches to the Evolution of Language: Social and Cognitive Bases* (Cambridge: Cambridge University Press, 1998), 2.

[74] Noam Chomsky, *The Minimalist Program* (Cambridge, MA: MIT Press, 1995).

anthropologists, Bruce Trigger and Robert Carneiro, prepared comprehensive and insightful reviews of this literature, yet left aside possible links to the literature on biological evolution.[75] A leading exception was Tim Ingold: his work, critical of Boasian relativism, reflected a broad and detailed reading of biological and social evolutionary literatures. He sought to combine the two frames by emphasizing mutually constitutive rather than interactive relations.[76] In another perspective on social evolution, sociologist Jonathan Turner completed multiple volumes, between 2000 and 2010, on the origin and impact of human emotions, a theory of human institutions, and the role of natural selection in social evolution.[77] These works propose a theory of societal evolution—at the macro-level and in the Holocene era—based "on institutions as a whole and as an emergent property of human social organization."[78] But the work of Turner, as with Ingold, continued to assume group behavior without clearly confirming the logic that distinguished it from individual behavior. In an iconoclastic overview, economist Graeme Snooks organized his analysis of the past two thousand years around the distinction between tactical and strategic institutions, where the latter determined the type of social strategy.[79] In the new millennium, anthropologist Stephen Sanderson published a 2007 survey of "an evolutionary interpretation of human society." Beginning with "classical" or Spencerian evolution, he traced anthropological, Boasian, Marxian, and sociological traditions, devoting

---

[75] Trigger, *Sociocultural Evolution*; Robert L. Carneiro, *Evolutionism and Cultural Anthropology: A Critical History* (Boulder: Westview Press, 2003). See also Robert L. Carneiro, "The Transition from Quantity to Quality: A Neglected Causal Mechanism in Accounting for Social Evolution," *PNAS* 97, 23 (2000): 12926–31. Later works include Jérôme Rousseau, *Rethinking Social Evolution: The Perspective from Middle-Range Societies* (Montreal: McGill-Queen's University Press, 2006); and John H. Miller and Scott E. Page, *Complex Adaptive Systems: An Introduction to Computational Models of Social Life* (Princeton: Princeton University Press, 2007).

[76] Tim Ingold, *Evolution and Social Life* (Cambridge: Cambridge University Press, 1986); In another approach, anthropologist C. R. Hallpike compared biological and social evolution, emphasizing that social structure adds to the complexity of social evolution. C. R. Hallpike, *Foundations of Primitive Thought* (Oxford: Clarendon Press, 1979); Hallpike, *The Principles of Social Evolution* (Oxford: Clarendon Press, 1986); Hallpike, *How We Got Here: From Bows and Arrows to the Space Age* (Central Milton Keynes: Authorhouse, 2008).

[77] Jonathan H. Turner, *Human Institutions: A Theory of Societal Evolution* (Lanham, MD: Rowman & Littlefield, 2003); Turner, *On the Origin of Societies by Natural Selection* (Boulder: Paradigm Publishers, 2008); Turner, *On the Origins of Human Emotions*; Turner, *Human Emotions*.

[78] "Institutional analysis is, therefore, inherently evolutionary because it explores how humans create population-wide structures and cultural systems that enable them to survive in the environment, often an environment of their own making." Turner, *Human Institutions*, 5.

[79] Snooks saw himself "exploding the myth of social evolution" in proposing a model of institutional change for the same two millennia. Graeme Donald Snooks, *The Ephemeral Civilization: Exploding the myth of social evolution* (London: Routledge, 1997).

a detailed chapter to "Evolutionary Biology and Social Evolutionism," and leading up to a summary of his theory of "evolutionary materialism."[80]

Building on the insights of Wallerstein's World-Systems theory, a Political Economy of World Systems section formed within the American Sociological Association in 1981; lively debates in historical sociology and international studies linked civilizational studies to political economy and social evolution.[81] Interest in long-term analysis grew inexorably among sociologists. In 1993, Barry Gills and Andre Gunder Frank edited *The World System: Five Hundred Years or Five Thousand?*, centering on a multidisciplinary research review that made the case for a five-millennium continuity in the main issues in political economy of large states and inter-state trade. In a somewhat parallel line of argument, political scientist David Wilkinson published on the notion of "central civilization," in which he traced a common envelope encompassing major states of Eurasia from the early days of Mesopotamia to the modern World-System.[82] In a major synthesis, Christopher Chase-Dunn and Thomas D. Hall published *Rise and Demise*. This work, incorporating a wide range of social-science literature, explored core-periphery relations in World-System analysis from 5000 years ago, including additional regions and World-Systems of smaller scale. The authors' explicit theory of evolutionary transformations relied on hierarchical networks of interconnections—of information at the widest scale, then of luxury-good exchange, political-military structures, and exchange of bulk goods in progressively narrower networks. The analysis centered on semiperipheral regions and their states as the "seedbed of change." It traced processes of growth and decline over time, giving primacy to interplay among units more than to dynamics within the various units.[83]

---

[80] Sanderson's evolutionary materialism takes the Neolithic revolution as the start of evolution and relies on demography, ecology, technology, and economy as causal factors. Individuals are units of selection and adaptation but the units of evolution are social groups and systems. In sum, the theory proposes categories of analysis but no specific mechanisms of change. The summary of the theory is followed by discussion of a helpful question, "In what way and when may social evolution be progressive?" Stephen Sanderson, *Evolutionism and Its Critics: Deconstructing and Reconstructing an Evolutionary Interpretation of Human Society* (Boulder, CO: Paradigm Publishers, 2007), 280–307.

[81] Most of the discussion centered on literate societies of the past five hundred years. The *Journal of World-Systems Research*, founded in 1995, was long edited by Christopher Chase-Dunn; Chase-Dunn directed the Institute for Research on World Systems (IROWS) at University of California—Riverside from 2002.

[82] Andre Gunder Frank and Barry K. Gills, "The Five Thousand Year World System: An Interdisciplinary Introduction," *Humboldt Journal of Social Relations* 18, 2 (1993): 1–79; David Wilkinson, "Central Civilization," *Comparative Civilizations Review* 17 (1987): 31–59. See also Gills and Frank, *The World System: Five Hundred Years or Five Thousand?* (London: Routledge, 1994).

[83] Christopher Chase-Dunn and Thomas D. Hall, *Rise and Demise: Comparing World-Systems* (Boulder, CO: Westview Press, 1997); for small-scale analysis, see Christopher Chase-Dunn, *The Wintu and the Neighbors: A Very Small World-System in Northern California* (Tucson: University of Arizona Press, 1998).

Historical studies expanded in scope from the 1990s. In an insightful work drawing on history of science, James Burke and Robert Ornstein wrote *The Axemaker's Gift*, a transhistorical critique of long-term and deep flaws in human culture.[84] At much the same time, the field of Big History arose through the innovative thinking of David Christian and Fred Spier, who each traced history from the Big Bang forward and gave significant attention to biological evolution, human evolution, and early human societies, relying on evidence from astronomy, physics, biology, thermodynamics, and studies of energy transfer, with a focus on "collective learning."[85] Through the co-authored work of J. R. McNeill and William H. McNeill, a concise and elegant volume, *The Human Web*, enabled world historians and the general public to consider the full extent of the human experience.[86] The field of world history became more fully organized in the 1990s: a journal appeared, programs of doctoral study gradually formed, and a debate about Eurocentrism put world-historical perspectives on the map.[87] The field of economic history spread beyond its North Atlantic concentration with comparisons of Europe and East Asia, and from that point took on comprehensive worldwide analysis, mostly for the modern period.[88] Studies of demography and historical demography, now aided by spreadsheets, expanded their attention to migration, social and labor history.[89] One potential benefit of the specificity of historical studies was the gradual design and creation of historical datasets. Extension of time frame, application of theory, and construction of data resources all characterized these multidirectional initiatives. A growing

---

[84] James Burke and Robert Ornstein, *The Axemaker's Gift: A Double-Edged History of Human Culture* (New York: Grosser/Putnam, 1995). Burke is a historian of science; the late Dr. Ornstein was a psychologist.

[85] In these works, "collective learning" is defined differently from "social learning," without reference to the dual-heritage literature. Christian, *Maps of Time: An Introduction to Big History* (Berkeley: University of California Press, 2003), 148–50; Spier, *Big History and the Future of Humanity*.

[86] J. R. McNeill and William H. McNeill, *The Human Web: A Bird's-Eye View of Human History* (New York: Norton, 2003).

[87] For an overview of world history intended to build institutions for its study, see Manning, *Navigating World History*.

[88] Kenneth Pomeranz, *The Great Divergence: China, Europe, and the Making of the Modern World Economy* (Princeton: Princeton University Press, 2000); Peter H. Lindert and Jeffrey G. Williamson, "Globalization and Inequality: A Long History," World Bank Annual Bank Conference on Development Economics—Europe (2001); Kevin H. O'Rourke and Jeffrey G. Williamson, *Globalization and History: The Evolution of a Nineteenth-Century Atlantic Economy* (Cambridge, MA: MIT Press, 1999).

[89] The International Institute of Social History (IISH), a major center of global and cross-disciplinary research, worked under Research Director Marcel van der Linden, supporting publication of large-scale social-historical studies in the *International Review of Social History*.

number of studies addressed multiple millennia, though with quite different methodologies and perspectives.[90]

Understandings of Evolution by 2010. Biological sciences consolidated the advances in theory and turned increasingly to applications. Environmental studies and Earth sciences became increasingly sophisticated in their handling of systems analysis, working with other disciplines. Cultural evolution spread beyond its initial focus on social learning to address other social issues at the individual level. Various visions of social evolution gathered unprecedented amounts of data and gained in clarity and distinctiveness. Scholars in each of these disciplinary domains undertook cross-disciplinary reading with growing seriousness.

Biological studies now advanced along both genetic and epigenetic paths. In genetics and genomics, focusing on DNA, specific and deterministic analyses proposed precise genetic evolution, estimating dates of genetic divergence, routes of migration, and specific biological characteristics. Human genomics gave strong indications on the migration of early humans throughout the globe, as described through studies of mitochondrial DNA and Y-chromosomes. Genetics now had a chronological dimension: while the efforts to document genomes in space and time led to some initial stumbles, results gradually became coherent and mutually reinforcing. These results tended to reconfirm that humanity is one large population, with considerable local diversity in its overall similarity: both individual differences and overall human biological equality were underscored as never before. In epigenetics, the revival of embryology and new studies in development of the organism led to an understanding that additional sorts of variation could take place through regulation of the expression of the genome.

Cultural evolution, a new conception, expanded at the border of biological and social science studies. Analysts of cultural evolution set the initiation of dual-heritage mechanisms at some time before the emergence of *Homo sapiens*, perhaps between 300,000 and 700,000 years ago. In research relying heavily on game theory, researchers proposed mechanisms by which genetic components affected social behavior or by which individual humans developed skills at imitation or social learning, retaining the results of their learning in their brains.

Cross-disciplinary analyses grew in this era, for instance in environmental and historical studies. Most prominently, successful creation of climate models made it possible to monitor the growing shifts in climate. The IPCC, under United Nations coordination, prepared detailed biennial reports

---

[90] Alfred W. Crosby, *The Columbian Exchange: Biological Consequences of 1492* (Westport, CT: Greenwood, 1972); Jared Diamond, *Guns, Germs, and Steel: The Fates of Human Societies* (New York: Norton, 1997); Spier, *Big History and the Future of Humanity*; Fred Spier, *The Structure of Big History from the Big Bang Until Today* (Amsterdam Univ. Press, 1996).

beginning in 1990. On a smaller scale, the expansion of graduate programs and academic journals in ecology, environmental studies, and environmental history created a multidisciplinary field that overlapped physical and biological fields and develop new initiatives in systems analysis.[91] Environmental and ecological studies now overlapped with geography, anthropology, and sociology, as reflected notably in the 2007 volume edited by ecologist Alf Hornborg and anthropologist Carole Crumley. The volume, including 21 contributions from 26 authors based in ten disciplines, opened many new questions in "socioenvironmental change" since the Neolithic.[92] At a wider level, historian David Christian and biologist Fred Spier each produced volumes in "Big History," setting the experience of human evolution in the context of larger-scale and longer-term history of the universe, the Earth, and life. Big History's wide conceptualization and valuable links rapidly made it a teaching field and then a source of guidance for researchers.

In "The Revolution that Wasn't," archaeologists Sally McBrearty and Alison Brooks made a firm and influential 2000 argument that human cultural innovations, rather than emerging belatedly at 40,000 years ago in Europe, had developed incrementally in Africa over a much longer time.[93] This essay arguably set the tone for a trend toward collaboration among social scientists in exploring long-term human history at the global level and to reading more widely across the disciplines as they did so. Studies in evolutionary linguistics and historical linguistics expanded in volume and sophistication. Migration studies, reaching across long periods of time, relied on textual data, linguistic data, and especially on the rapidly expanding genetic data on migration. Sociologists, working in the World-Systems framework, expanded collaborations with political scientists, geographers, anthropologists, and ecologists. The field of world history, still small in comparison with other historical subfields, underwent dramatic expansion both in teaching and research. While the concept of civilization still played a central role in many world-historical studies, other world historians read more widely and extended their time frame to the era before the rise of cities. Put in other terms, there were now overlaps between studies of world history and Big History.

---

[91] For an outstanding example, see John R. McNeill, *Something New Under the Sun: An Environmental History of the Twentieth-Century World* (New York: W. W. Norton, 2000).

[92] Alf Hornborg and Carole L. Crumley, eds., *The World System and the Earth System: Global Socioenvironmental Change and Sustainability Since the Neolithic* (New York: Routledge, 2016 [2007]). Also studies in environmental history, as by J. R. McNeill.

[93] Sally McBrearty and Alison Brooks, "The Revolution That Wasn't: A New Interpretation of the Origin of Modern Human Behavior," *Journal of Human Evolution* 39 (2000): 453–563.

CHAPTER 12

Cross-Disciplinary Analysis in Global Tension, 2010–2020

The second decade of the twenty-first century brought expanded tensions, worldwide, at several levels. A global financial crisis in 2008–2009 was halted short of economic collapse only through governmental spending of huge amounts of public funds to prop up collapsing banks and industries. Environmental degradation expanded on land and in the waters, yet governments took virtually no action despite periodic warnings arising from IPCC reports: floods, fires, and flows of refugees accelerated as temperatures continued to rise. Tension arose in global politics as more nations became globally influential. China's Gross Domestic Product surpassed that of the United States by some measures; India, Brazil, Indonesia, Nigeria, Philippines, and Pakistan became influential along with EU nations, Russia, and Japan, but in an atmosphere of growing national isolation rather than collaboration.

Little increased income reached common people, though they did benefit from innovations such as mobile phones and social media. Notably in Africa, where communications had long been limited for lack of investment in landlines, the arrival of mobile phones and the inexpensive construction of mobile towers allowed the continent's personal communication to equal the worldwide level. Further, links within a global system of universities became increasingly integrated, yet with limits on movements of students. Growing levels of popular awareness of multiple scales in time and levels of aggregation enabled the notion of a shared Human System, in which all participated, to become more plausible.

Expanded study in biological evolution. Genetic analysis and paleontology each continued the advances of the previous period. Most spectacularly, advances in genome sequencing combined with advances in techniques of retrieving DNA from fragments of ancient human remains. Thus, the study of "ancient DNA" became well-funded, focusing especially on three research groups, led by Svante Pääbo at the University of Leipzig, Eske Willerslev at the University of Copenhagen, and David Reich at Harvard University. One

© The Author(s) 2020
P. Manning, *Methods for Human History*,
https://doi.org/10.1007/978-3-030-53882-8_12

advantage of the new technique was that it often permitted sequencing of the entire genome of the subject, thus giving a wider range of results in contrast to previous work, which had focused on restricted portions of the genome. The other advantage was that, by sequencing genomes of persons who lived several thousand years ago or more, one could obtain a look at samples from a regional population before it was overlaid by recent migrations, making it easier to trace distant ancestry. A dramatic result of ancient DNA studies was the discovery of the hominid population labeled Denisovan, initially from a portion of a finger bone from the Denisova cave of Kazakhstan. This genetic result led soon to the understanding that a Denisovan population had existed in much of eastern and perhaps southern Asia, and that results of its interbreeding showed up especially in populations of Melanesia.[1]

The early years of genomic study had left Africa in relative neglect, though it was evident that the greatest diversity was to be found among the people of the home continent. New initiatives in genetic and paleontological research opened up in early years of the new century, and by 2010, results were appearing in publications, led especially by Sarah Tishkoff.[2] Meanwhile, researchers in paleontology continued to discover early human remains in Africa and Asia. Expanded genetic research in Africa, of interest in itself, encouraged further paleontological research that advanced results in both fields. With that, efforts were redoubled to locate ancient DNA in Africa, despite the difficulties of its preservation. In a remarkable set of results from the Shum Laka site in southwestern Cameroon, DNA data were retrieved from the remains of four children who had lived from 8000 to 3000 years ago.[3] For the time in which the children were alive, the results show some hints of the migration of Bantu-speaking people into this area as they moved to the east and south. For much earlier times, these genomes revealed new complexities and differences in the African populations that existed long before the expansion of speaking peoples some 70,000 years ago. David Reich describes other research in which people of both the Yoruba ethnicity of West Africa and the San of Southern Africa are shown to have descended from "a mixture of two highly differentiated human populations" that took

[1] Reich, *Who We Are*, 53–62.

[2] Tattersall, *Masters of the Planet*. In a sign of changing times, in 2010 *Current Biology* published a set of continental reviews of genetic variation, in which African evidence was addressed in a very detailed study. See Michael C. Campbell and Sarah A. Tishkoff, "The Evolution of Human Genetic and Phenotypic Variation in Africa," *Current Biology* 20 (2010): R166–R173. See also Jibril Hirbo, A. Ranciaro, and S. A. Tishkoff, "Population Structure and Migration in Africa: Correlations Between Archaeological, Linguistic and Genetic Data," in *Causes and Consequences of Human Migration: An Evolutionary Perspective*, C.B.C. (2009): 135–71. https://doi.org/10.1017/cbo9781139003308.011. At a more general level, see Jonathan K. Pritchard, Joseph K. Pickrell, and Graham Coop, "The Genetics of Human Adaptation: Hard Sweeps, Soft Sweeps, and Polygenic Adaptation," *Current Biology* 20 (2010): R206–R215.

[3] Mark Lipson, Isabelle Ribot, David Reich, et al., "Ancient West African Foragers in the Context of African Population History," *Nature* 577 (2020): 665–70.

place over two hundred thousand years ago. This suggests the depth and variety of human ancestry in Africa.[4]

Language. Advances and debates in study of evolutionary linguistics were especially lively, though the debates were not rapidly resolved.[5] W. Tecumseh Fitch, in a 2010 synthesis aimed at explaining the work of evolutionary biologists and linguists to each other, summarized both fields convincingly.[6] His approach reaffirmed an opening to various sorts of "protolanguage" that may have existed, focusing on gesture (as analyzed by Tomasello) or emphases on music and dance.[7] Fitch summarized but left unresolved the choice among four hypotheses for the evolution of syntax.[8] Yet Fitch's work, otherwise comprehensive, set aside the issue of language universals, for which a collection had appeared in 2009: this volume confirmed continuing advances in study of language universals and indicated their relevance to evolutionary linguistics.[9] Fitch also chose not to explore the expanding research campaign in cultural evolution.

Then Berwick and Chomsky, in a 2016 book, linked the points that each of them had previously introduced.[10] They proposed a Basic Property of language, with three elements that apparently emerged in a certain order.[11] First was that "a conceptual system for inference, interpretation, planning, and the organization of action—what is informally called 'thought'"—that had built

---

[4] In this view, the separations leading to Yoruba and San genomes took place in different paths. Further, "An expansion of modern humans ... after around fifty thousand years ago could then have connected all populations of Africa." Reich, *Who We Are*, 209–13.

[5] As an example of the continuing debates, in this case over the physical basis of human speech, see the article and forum in Philip Lieberman, "The Evolution of Human Speech: Its Anatomical and Neural Bases," *Current Anthropology* 18 (2007): 39–66.

[6] To clarify his wide-ranging discussion, Fitch distinguishes "the faculty of language in a broad sense" (FLB, referring to communication among all species) from "the faculty of language in a narrow sense" (FLN, referring uniquely to humans and spoken language). Fitch, *Evolution of Language*, 21–23.

[7] Fitch, *Evolution of Language*, 401–507; Michael Tomasello, *Origins of Human Communication* (Cambridge, MA: MIT Press, 2008), 57–71; William H. McNeill, *Keeping Together in Time: Dance and Drill in Human History* (Cambridge, MA: Harvard University Press, 1995).

[8] Fitch's four choices are: syntax arose through cultural processes; Universal Grammar arose through natural selection; syntax originated in social knowledge; and syntax originated in visual processing. Fitch, *Evolution of Language*.

[9] Barbara Finlay, from the standpoint of the expanding evo-devo field, suggests a possible interaction between language evolution and language universals. Barbara L. Finlay, "Evolution, Development, and Emerging Universals," in Morten H. Christiansen, Chris Collins, and Shimon Edelman, eds., *Language Universals*, 261–65 (Oxford: Oxford University Press, 2009).

[10] Robert C. Berwick and Noam Chomsky, *Why Only Us: Language and Evolution* (Cambridge, MA: MIT Press, 2016).

[11] Berwick and Chomsky, *Why Only Us*, 10. A contrasting approach by Steven Pinker assumes that language is primarily to serve the function of communication, and that it evolved in a gradual fashion. Pinker, *The Language Instinct: How the Mind Creates Language* (New York: Harper, 2007). A remaining puzzle is how to have a lexicon without speech.

up over time. Second was a process of logical operation, labeled "Merge" in the earlier work of Berwick. This mutation had made it possible to organize the elements of the system of thought so as to create hierarchical thinking: the process linked any two concepts together, then treated the two concepts as a unit and added another concept to it, repeating the operation by adding another element, until by stages a complex set of statements emerged. Third and last of the steps was the creation of speech—a process for translating the now-advanced thinking into messages to be sent beyond the individual, and also for receiving and interpreting messages from others. The Merge characteristic, a genetic mutation, would surely have spread broadly in the human population, as it enabled solving an individual's problems at a higher level. In that case, the creation of verbal communication was an entirely separate step, coming well after the spread of higher-order conceptualization. This argument—that language is about higher-order thought more than about communication and that human language is greatly distinct from any forms of animal communication—is consistent with the sudden appearance and rapid expansion of spoken language. For the Merge thesis in linguistics, one research group has sought to confirm it through activity observed in a certain brain area.[12]

Emotions. Studies of emotion had gradually expanded from the 1970s in the fields of biology, psychology, sociology, and history.[13] In 2018, two impressive reports on research appeared, based on quite different research projects, but adding a great deal to the current understanding and further questions on emotions. Adolphs and Anderson, in *The Emotions in Neuroscience*, emphasized their own theory of emotional states but also gave careful reviews of the topics and theories of many other studies.[14] They identified a set of core *emotional states*, assumed by many to be inherited and largely shared by all animals. These emotional states are surrounded by stimuli that elicit them and other elements of body states; within this level, there may be responses of emotions to stimuli, as with fear, described in terms of arousal and valence.[15] The second step was to categorize the experience of emotion, the feelings of emotions in humans and perhaps in animals. Conscious experience of an emotion ("feelings" or "affect") depends on content that is somatic (body), perceptual (world), and cognitive (mind);

---

[12] The specificity of the Merge hypothesis has been rewarded with research results claiming to identify the function and its location in the brains of contemporary subjects in Germany. Emiliano Zaccarella and Angela D. Friederici, "Merge in the Human Brain: A Sub-Region Based on Functional Investigation in the Left Pars Opercularis," *Frontiers in Psychology* 6 (2015): 1818.

[13] See also Chapter 7.

[14] Ralph Adolphs and David J. Anderson, *The Neuroscience of Emotion: A New Synthesis* (Princeton: Princeton University Press, 2018).

[15] The emotion state is assumed to have motivational, behavioral, and functional aspects of approach/withdrawal. Adolphs and Anderson accept the idea of a taxonomy of perhaps six basic emotions but decline to propose a list until it can be sustained by research distinguishing them.

the latter includes expectations about what might happen in the future or did happen in the past. Third, Adolphs and Anderson observe that most theories of emotional experience have a layered architecture, which should add further elements to the understanding of conscious experience. Within this framework, Adolphs and Anderson focused on analyzing emotion states, addressing many species.

Psychologist Lisa Feldman Barrett, in contrast, focused on analyzing the experience of emotions among humans: she viewed emotions as constructed out of human social experience, rather than conveyed by biology.[16] She argued that infants, in learning to understand speech and to speak, adopt concepts of emotions that guide them in interactions with others. The discourse on emotions is thus different from the emotions themselves, thus complicating the task of research—and complicating the behavior of speaking humans. Barrett and the team of Adolphs and Anderson, in their wide-ranging analyses, nevertheless leave out the substantial work on emotion published by historians. A survey of historians' analysis focuses primarily on the experience of emotions, dividing the approaches into emotionology, emotional regimes, emotional communities, and emotions as performances.[17] Finally, a collection edited by von Scheve and Salmela gives concise explorations of collective emotions at multiple scales.[18]

Cultural evolution. A 2017 review essay on cultural evolution by Nicole Creanza, Oren Kolodny, and Marcus Feldman included both new voices and old among the theorists of cultural evolution. It made the case that the framework of cultural evolution has sustained its coherence and has extended its influence in evolutionary studies, relying on the standpoint of change for the individual human organism.[19] In this view, a broad consensus arose among cultural-evolution researchers at this scale, although the constituent subfields maintained their specificity. The boundaries surrounding and within this expanding field are not yet clear. For instance, dual-heritage analyses gave little attention to environmental changes, to group-level cultural issues, or to research in social evolution; in addition, Creanza et al. left aside the

---

[16] Barrett, *How Emotions Are Made*. Joseph Lopreato's 1984 book on "behavioral predispositions" in humans, though inspired by E. O. Wilson's vision of sociobiology, addressed the experience of emotions more than emotion states. Lopreato, *Human Nature and Biocultural Evolution*. See also Jonathan H. Turner, *On the Origins of Human Emotions: A Sociological Inquiry into the Evolution of Human Affect* (Stanford: Stanford University Press, 2000); Jonathan H. Turner, *Human Emotions: A Sociological Theory* (London: Routledge, 2007).

[17] Rosenwein and Cristiani note the founding work of Peter Stearns in history of emotions. Barbara H. Rosenwein and Riccardo Cristiani, *What Is the History of Emotions?* (Cambridge: Polity, 2018), 29–49. See also Nicole Eustace, et al., "*AHR* Conversation: The Historical Study of Emotions," *American Historical Review* 117 (2012): 1487–531.

[18] Christian von Scheve and Mikko Salmela, eds., *Collective Emotions: Perspectives from Psychology, Philosophy, and Sociology* (Oxford: Oxford University Press, 2014).

[19] Nicole Creanza, Oren Kolodny, and Marcus W. Feldman, "Cultural Evolutionary Theory: How Culture Evolves and Why It Matters," *PNAS* 114, 30 (2017): 7782–89.

question of whether the expanded analysis of cultural evolution would also interact with analysis of ontogeny or social evolution.[20] Nevertheless, the implicit argument of their approach is that additional areas of study—including visual communication, evolutionary linguistics, agricultural innovation, emotional states, paleontology, and even the circulation of memes—may be seen as encompassed by the broad net of cultural evolution.[21] In one sense, these emerging subfields fit within E. O. Wilson's claim that sociobiology was the new synthesis for biology; in another sense, most of these advances have proceeded quite independently of Wilson's approach.[22]

Among the many examples that can be drawn from the widening range of cultural-evolution studies is the volume edited by Alex Mesoudi and Kenichi Aoki, who applied the cultural evolution paradigm to Neanderthal evidence.[23] Kevin Laland, in an overview of the cultural evolution paradigm and the many topics into which it has reached, emphasized that copying behavior, the core element of cultural evolution, is ubiquitous throughout the history of vertebrates and that its consequences can be traced from the level of fishes to the highest level of the human arts.[24] In the same volume, Laland provided a detailed interpretation of the rise of spoken language that relied heavily on the work of Bickerton and mechanisms of cultural evolution. This interpretation gave considerable attention to protolanguage and allowed for the possibility that syntactic speech was a relatively sudden invention.[25] In another new direction that is also allied to inclusive fitness, Tim Lewens argued that "selectionist" thinking in cultural evolution has commonly and wisely been replaced with "kinetic theories of culture,"

---

[20] Boyd and Richerson, *Origin and Evolution of Cultures*; Creanza et al., "Cultural Evolutionary Theory."

[21] Examples of major interpretations that may be included within this expanded framework include the work of Ian Tattersall in paleontology and Michael Tomasello in evolutionary psychology. Tattersall, *Masters of the Planet*, and Michael Tomasello, *Becoming Human: A Theory of Ontogeny* (Cambridge, MA: Belknap Press of Harvard University Press, 2019).

[22] For instance, dual-heritage scholars were able to argue early on that their mechanisms and hypotheses differed from those advanced by sociobiologists, so that the differences were testable. Boyd and Richerson, *Culture and the Evolutionary Process*, 199. On the challenge of E. O. Wilson and colleagues to the validity of inclusive fitness is a category of analysis, see Nowak, et al., "The Evolution of Eusociality"; Abbott, et al., "Inclusive Fitness Theory and Eusociality"; and Wilson, *Sociobiology*.

[23] Alex Mesoudi and Kenichi Aoki, eds., *Learning Strategies and Cultural Evolution in the Palaeolithic* (Tokyo and Heidelberg: Springer, 2015); Mesoudi, *Cultural Evolution*. See also J. Henrich, *The Secret of Our Success: How Culture Is Driving Human Evolution, Domesticating Our Species, and Making Us Smarter* (Princeton: Princeton University Press, 2016).

[24] Kevin N. Laland, *Darwin's Unfinished Symphony: How Culture Made the Human Mind* (Princeton: Princeton University Press, 2017).

[25] The chapter, focused on forms of communication, makes an effective case that advances in communication took place among early hominin, but does little to address the distinctiveness of syntactic language. Laland, *Darwin's Unfinished Symphony*, 175–207.

which rely on a general form of population thinking.[26] Richard McElreath, a leader in dual-heritage analysis, came to direct a research program at the Max Planck Institute for Evolutionary Anthropology.[27] Peter Turchin, whose initial work was in population ecology, turned to applying his demographic skills to issues in both social history and cultural evolution. In an explicit extension of cultural evolutionary theory, Turchin focused on levels of violence in human communities, hypothesizing changing levels of violence from the Neolithic era, 10,000 years ago.[28] In developments further afield advances in nuclear-magnetic-resonance technology made it possible to trace the folded shapes of proteins and to unveil processes of epigenetics, which regulated activity and inactivity of genes, and by which elements of phenotypical heredity are passed on without direct passage through DNA. By this mechanism, biological evolution could still be going on continuously within *Homo sapiens*, even as genes changed only slowly. If this trajectory persists, future research in cultural evolution will need to account for many distinctive processes and analytical subfields within the arena of cultural evolution, as well as coevolution with processes of biological and social evolution.

Large-scale social change. Social scientists have recently been active in expanding the spatial and temporal scope of their disciplinary analysis, venturing to combine study of the past five hundred years with earlier aspects of human heritage—some supporting the general notion of social evolution, others restricting themselves to more specific analyses. Several studies have traced violence over the course of recorded history, developing single-thesis interpretations of the human experience of violence.[29] Evolutionary

---

[26] Tim Lewens, *Cultural Evolution: Conceptual Challenges* (Oxford: Oxford University Press, 2015). The approach of Lewens is rather close to the sociobiology of E. O. Wilson. For critiques of Lewens, see Wybo Houkes, "Population Thinking and Natural Selection in Dual-Inheritance Theory," *Biological Philosophy* 27 (2012): 401–17; and Peter J. Richerson, "Recent Critiques of Dual Inheritance Theory," *Evolutionary Studies in Imaginative Culture* 1 (2017): 203–11. Does "kinetic" refer to EO Wilson's framework?

[27] Richard McElreath, "A Long-form Research Program in Human Behavior, Ecology and Culture" (2017).

[28] Peter Turchin, *Ultra Society*.

[29] Douglass C. North led a group analyzing "limited access orders," "open access orders," and their relationship to levels of violence. Walter Scheidel traced recurring narratives of rise of inequality in societies over time, followed by periodic and violent leveling of inequality. See Douglass C. North, John Joseph Wallis, Steven B. Webb, and Barry R. Weingast, *Violence and Social Orders: A Conceptual Framework for Interpreting Recorded History* (New York: Cambridge University Press, 2009); Douglass C. North, John Joseph Wallis, Steven B. Webb, and Barry R. Weingast, "Limited Access Orders: An Introduction to the Conceptual Framework," in Douglass C. North, John Joseph Wallis, Steven B. Webb, and Barry R. Weingast, eds., *In the Shadow of Violence: Politics, Economics, and the Problems of Development* (New York: Cambridge University Press, 2013), 1–23; Walter Scheidel, *The Great Leveler: Violence and the History of Inequality from the Stone Age to the Twenty-first Century* (Princeton: Princeton University Press, 2017). See also Eric Alston, et al., *Institutional and Organizational Analysis: Concepts and Applications* (Cambridge: Cambridge University Press, 2018).

psychologist Stephen Pinker also wrote on the history of violence, arguing according to a different theory that violence has declined steadily in the course of the Holocene era.[30] These detailed studies will ultimately advance the analysis of emotions and behavior. Nevertheless, the decisions by researchers to limit their studies to a single emotion implicitly assume that emotions are independent of one another, in contrast to the interplay assumed by specialists in emotions.

Anthropologists Kent Flannery and Joyce Marcus advanced the argument that communities of late Pleistocene and early Holocene times valued social equality and developed numerous techniques to sustain equal social relations in response to the growing pressures for hierarchy.[31] Steven Mithen had earlier focused on a similar time frame, beginning well before the rise of agriculture, to demonstrate the responses of communities worldwide to the warming aftermath of the Glacial Maximum.[32] In a 2013 collective volume, co-editors Tim Ingold and Gisli Palsson and nine other anthropologists offered perspectives on the integration of anthropology and life sciences, though not yet with a research agenda.[33]

The World-Systems perspective on social change, with an institutional base at the Institute for Research on World-Systems (IROWS) and a wide-ranging network, reaffirmed its approach through formation of an interdisciplinary working group on settlements and polities, comparing World-Systems.[34] Earlier, Christopher Chase-Dunn, director of IROWS, had paired with psychologist Bruce Lerro to publish a 2014 general overview of human evolution, focusing analysis on growth and conflicts in systems transformed by semiperipheral initiatives, but also surveying biological evolution and progress.[35] The prolific sociologist Jonathan Turner, who has

---

[30] Steven Pinker, *The Better Angels of Our Nature: Why Violence Has Declined* (New York: Viking, 2011).

[31] Flannery and Marcus, *The Creation of Inequality*.

[32] Steven Mithen, *After the Ice: A Global Human History, 20,000–5000 BC* (Cambridge, MA: Harvard University Press, 2003).

[33] Tim Ingold and Gisli Paisson, eds., *Biosocial Becomings: Integrating Social and Biological Anthropology* (Cambridge: Cambridge University Press, 2013).

[34] Christopher Chase-Dunn, et al., "The SETPOL Framework: Settlements and Polities in World-Systems." IROWS Working Paper #114, irows.ur.edu/papers/irows114.irows114.htm; Elizabeth Bogumil and Christopher Chase-Dunn, "Settlement Networks and Sociocultural Evolution," IROWS Working Paper #127, irows.ucr.edu/papers/irows127/irows127.htm.

[35] Christopher Chase-Dunn and Bruce Lerro, *Social Change: Globalization from the Stone Age to the Present* (Boulder: Paradigm Publishers, 2014); Shryock and Smail, *Deep History*; Ian Morris, *Why the West Rules—For Now: The Patterns of History, and What They Reveal About the Future* (New York: Farar, Straus, and Giroux, 2010); Morris, *The Measure of Civilization: How Social Development Decides the Fate of Nations* (Princeton, NJ: Princeton University Press, 2013); Peter Turchin, *Ultra Society*; Leonid Grinin, Andrey V. Korotayev, Robert L. Carneiro, and Fred Spier, *Evolution: Cosmic, Biological, Social* (Volgograd: Uchitel, 2011).

collaborated with the World-Systems group, joined with Richard Machalek to make more explicit the efforts to link sociological analysis to biological reasoning.[36]

Scholars in historical disciplines produced a growing number of major interpretations, creating or relying on large-scale datasets to support their analysis. John L. Brooke provided a substantial review of climate change and its influence in the Earth system and throughout human history.[37] Historian Ian Morris published a sweeping two-volume analysis of the period from the rise of civilization to the present. Morris sought to quantify "social development" as a macro-level summary of social achievement.[38] Historical studies gained institutional strength, notably as the long-established International Institute of Social History (IISH) in Amsterdam expanded its temporal and disciplinary reach in the twenty-first century. Housed at IISH, CLIO-INFRA formed in 2010 to build datasets on global social, economic, and institutional history since 1500. Also housed at IISH was the Global Collaboratory on the History of Labour Relations. Among the many books published in the wake of these data collections, Pim de Zwart and Jan Luiten van Zanden published a multi-regional survey of economic and social globalization since 1500.[39] At a longer time frame, world and global history expanded at both the levels of general readership and academia. In 2015, Yuval Noah Harari's *Sapiens* became a best-seller in multiple languages; in the same year, the ten volumes of the *Cambridge World* History appeared, addressing the full scope of human history.[40]

---

[36] Jonathan H. Turner and Richard S. Machalek, *The New Evolutionary Sociology: Recent and Revitalized Theoretical and Methodological Approaches* (New York: Routledge, 2018).

[37] Brooke, *Climate Change and the Course of Global History: A Rough Journey* (Cambridge: Cambridge University Press, 2014).

[38] Morris developed index figures for numerous civilizational regions for much of the Holocene era, measuring their relative success in terms of the amount of energy they were able to capture and consume: in this approach, he sought to follow the lead of Leslie White. Ian Morris, *Why the West Rules—For Now: The Patterns of History, and What They Reveal About the Future* (New York: Farar, Straus, and Giroux, 2010); Morris, *The Measure of Civilization: How Social Development Decides the Fate of Nations* (Princeton, NJ: Princeton University Press, 2013).

[39] CLIO-INFRA (www.clio-infra.eu); Global Collaboratory on the History of Labour Relations (https://collab.iisg.nl/web/LabourRelations). For a wide-ranging review of socio-economic history by the IISH research director, see Leo Lucassen, "Working Together: New Directions in Global Labour History," *Journal of Global History* 11 (2016): 66–87; see also Pim de Zwart and Jan Luiten van Zanden, *The Origins of Globalization: World Trade in the Making of the Global Economy, 1500–1800* (Cambridge: Cambridge University Press, 2018. For a historian's argument for expanded, multidisciplinary analysis of issues in inequality and environmental degradation, see Patrick Manning, "Inequality: Historical and Disciplinary Approaches," *American Historical Review* 122 (2017): 1–22. These projects, while limited to the past millennium, show a new coherence and comprehensiveness in the research agenda of historians.

[40] Yuval Noah Harari, *Sapiens: A Brief History of Humankind* (New York: Harper, 2015); Merry Wiesner-Hanks, general editor, *The Cambridge World History*, 10 vols. (Cambridge: Cambridge University Press, 2015).

In an important consolidation of approaches to knowledge of natural and social sciences, in 2018 UNESCO announced the formation of the International Science Council (ISC), through the merger of the International Council for Science (ICSU) and the International Social Science Council (ISSC). This restructuring, while initially at the level of global administration, reflected growing cross-disciplinary ties and expanded the hope of overcoming the long isolation of social sciences from the natural sciences.

Group behavior and institutional evolution. Further exploration of the collective intentionality framework among philosophers led to the articulation of a theory of group behavior. Raimo Tuomela led in this work, ultimately developing a coherent theory of group behavior and institutional functioning. He drew on the work of economist Michael Bacharach, who had made important strides toward formulating game theory in a way that allowed for team behavior.[41] By 2007, these efforts were combined, and in 2013, Tuomela published a book presenting a fairly complete theory of group behavior. Tuomela, in articulating the functioning of social groups, hypothesized a shift of agency from individual to collective and a shift of reasoning from I-reasoning to we-reasoning—a shift that can also be described as from "implicit cooperation" to "explicit cooperation."[42] As he argued, once one agrees to join with others in we-mode, collective reasoning becomes computationally simpler by reducing the number of alternatives, and it reduces the amount one needs to know about others' thinking to decide on actions. We-mode thinking is better able to handle collective action dilemmas, as "What should I do?" becomes "What should we do as a group?" It enables creation of more social order and gives better explanations of society. To unify a group, one requires a group reason (a unifying reason for group members to participate in group-based activities), a collectivity condition for all members ("necessarily being in the same boat"), and collective commitment (basically a product of joint intention and the members' group reason).[43] Tuomela sought to explore the cultural evolution of cooperative social activities by linking his framework to the analysis of Boyd and Richerson, expressing interest in their concept of "ultrasociality," but had to give up on that approach because he found that their framework did not allow for "the creation of cultural and social institutions by means of performative collective acceptance."[44]

Most importantly, Tuomela asserted that the we-mode is conceptually *irreducible* to the I-mode: the solutions to collective action problems in we-mode

---

[41] Bacharach died suddenly in 2002; his draft book was published posthumously. Michael Bacharach, "Interactive Team Reasoning: A Contribution to the Theory of Cooperation," *Research in Economics* 53 (1999): 117–47; Bacharach, eds. N. Gold and R. Sugden, *Beyond Individual Choice: Teams and Frames in Game Theory* (Princeton: Princeton University Press).

[42] See Chapter 5, Topic 5.2.

[43] Tuomela, *Social Ontology*, 6–7.

[44] Tuomela, *The Philosophy of Sociality: The Shared Point of View* (New York: Oxford University Press, 2007), 227, 231.

are different, smaller in number, and more advantageous to the group than the solutions generated in the I-mode. He defended this assertion through a team-based game theory exercise with the Hi-Lo game. He argued that, "the social world can be adequately understood and rationally explained only with the help of we-mode concepts in addition to I-mode concepts."[45]

My own intervention in this analysis has been to apply Tuomela's theses to the issue of human social evolution, by assuming that creation and reproduction of social institutions is the core element of social evolution. The theory of social change through institutional evolution arose out of an attempt by this author to interpret long-term human history, with close attention to the transition to the use of spoken language by humans in Africa some 70,000 years ago.[46] While the timing, location, and causation of the emergence of spoken language cannot yet be verified, an immense amount of circumstantial evidence points in that direction. Articulation of the theory of institutional evolution adds even more evidence to this case. While processes of cultural evolution, governed by a dual-heritage mechanism, gradually facilitated growth in human skills and cooperation both in earlier times and later times, the emergence of spoken language brought a sharp break. The invention of spoken language required the creation of much larger human communities, including perhaps 150 persons. While individual agency was required to join the speaking community—especially the great effort of learning the vocabulary and syntax—the language could only be sustained through group collaboration. The speaking community, therefore, became the first social institution. Its creation launched the process of social evolution. The theory of social evolution is thus both deductive and historical. The dynamics proposed for institutional evolution are largely Darwinian. Institutional evolution required the construction of institutions by members who chose to join and support them, working with the institutions through *variation* (in innovations), *inheritance* (nature of the social archive and its reproduction), *selection* (at levels of local and societal functionality), *fitness* (as measured by reproduction or social welfare; and assessment by beneficiaries at levels from institutional leaders to society generally).

Application of the theory to institutional history naturally encounters complications. In academic parlance, the term "institution" is defined in many different ways. Another complication is the realization that cultural evolution and institutional evolution make opposite predictions about diversity within communities: institutional evolution thrives on diversity while cultural evolution is weakened by diversity. A further complication is that institutional evolution assumes that beneficiaries of an institution will periodically review

---

[45]. Tuomela's verification of the nonreducibility of we-mode thinking to I-mode thinking was not a repudiation of the 1965 analysis by G. C. Williams, which argued that the "group evolution" thesis of Wynne-Edwards was not sustainable and reduced in practice to individual-level behavior. Instead, Tuomela's identification of both the character of the group and the standard for its fitness is different from those assumed by Wynne-Edwards. See Tuomela, *Social Ontology*, 7, 179–213; Williams, *Adaptation and Natural Selection*.

[46] Manning, *History of Humanity*, 36–61.

it, reforming it, or even closing it as needed. In practice, the core leaders of each institution tend to direct it, and the process of selection appears to work poorly.[47] So far, the theory draws attention to this apparent malfunction of institutional evolution but does not resolve it. Nevertheless, the theory has the advantage of encouraging detailed description and analysis of social institutions, suggesting specific dynamics that characterize various institutions.

When did social evolution begin? As the twenty-first century proceeded, evidence in numerous fields contributed to the argument that human speech began, rather rapidly, about 70,000 years ago. While other important transitions retain their importance, adding this change to the existing list makes a major difference in the overall understanding of human evolution. Before speech came bipedalism, the current human skeletal form, big brains, and social learning; after speech came the complex processes summarized as migration, agriculture, cities, and industry. But decisions on prioritizing these transitions do much to determine the character of our interpretation of human evolution. To assume the social evolution of all speaking humans since 70,000 years ago is to adopt an *inclusive* approach—it emphasizes human diversity, in that all people today are descended from the initial community.[48] Analyses of social evolution of city-dwellers in the last 5000 years, and of agriculturists in the last 10,000 years, take *exclusive* approaches to human evolution, leaving rural and non-agricultural populations out of the story. Further, the creation of language and of speaking communities was arguably a transition of immense importance, in which the shared work of preserving that first institution set the model for many succeeding institutions.

Many of the arguments supporting rapid emergence of spoken language some 70,000 years ago have come to light only recently. Others have been available for many decades. The McBrearty-Brooks article of 2000 emphasized the long-term continuity in human cultural innovation, beyond 100,000 years ago. Recent work in evolutionary linguistics, while it does not provide a date for syntactic speech, shows the many preconditions for speech yet also shows that spoken language is quite different from other social activities of humans or other species. The study of historical linguistics traces the path of human migration outward from Northeast Africa. Genomic analyses confirm this migration pattern and date its initiation with even more detail. Migration paths show the importance of maritime migration—and therefore of maritime technology—both at the start of the global diaspora and at several stages along the way. The brilliant cave paintings at Lascaux, France, discovered in 1940 and now dated to 15,000 years ago, were initially conceived

---

[47] James Coleman's view of social theory, foundational to rational choice theory, includes a somewhat parallel argument in which he calls for a review of the "micro – to – macro problem," "from the purposive action of individual actors to the functioning of a system of action." Coleman, "Social Theory," 1320–27.

[48] Of course, this approach leaves out non-speaking hominids, though it assumes that many of them were absorbed into the expanding, speaking population.

as unique accomplishments. They remain unique, but recent discovery of impressive cave paintings in Sulawesi and Borneo, of 40–44,000 years of age, add to the record of known ancient art work of extraordinary skill from India, Australia, Africa, Siberia, and elsewhere in Europe. These works emphasize that the early achievements of speaking humans were especially in representation—visual but surely also conceptual—and that technology came more gradually. This insight is consistent with the systemic analysis of James G. Miller, who argued in 1978 that living subsystems dealing with information were more numerous than those addressing matter and energy. Further, the analyses of hominin group sizes by Dunbar and Aiello reinforce Tuomela's vision of collective intentionality, implying that human group sizes increased sharply once speech began. Overall, these arguments made more plausible the exploration of institutional evolution within communities of speaking humans as a mechanism for social evolution, beginning about 70,000 years ago.

Understandings of evolution by 2020. In this book, I have sought to use the term "human evolution" as an inclusive term, encompassing change in the human species over varying time periods, at multiple scales of aggregation, and with multiple processes of change. Within that broad definition, the purpose of the book has been to introduce numerous methods for study of human evolution at various scales in Part I and to trace the development of knowledge of human evolution at the same range of scales in Part II. Here is a summary of six scales of contemporary research—from the individual scale up to the global scale—indicating the disciplines that are now active at each scale, the topics and methods of their work. I have indicated disciplines that work with Darwinian logic, in whole (in biology) or in part (in all the rest). The six scales themselves are listed in temporal order of the time frames to which they apply. For the discussion *within* each of these scales, the disciplines are listed in chronological order of their intervention in interpretation, from the earliest enunciated to most recent. Thus, the narrative framework of civilizations was the first to address civilizational groupings; it has faded in importance but is still applied by some researchers.

Scale 1: genetic and epigenetic composition of individuals.
Time frame: millions of years.
Disciplines:

- *anthropology and paleontology*: study of skeletal remains and archaeological surroundings of early hominids;
- *biology* [Darwinian]: analyze genomes and life-course development in individual humans. Most recently, whole-genome analysis, for modern or ancient DNA, provides histories of ancestry from the short term to hundreds of thousands of years ago. Biology is also applied to populations at all the larger scales, treating them as aggregations of individuals.

Scale 2: individual genome as connected to social practices.
Time frame: since before *Homo sapiens*.

Disciplines:

- *cultural evolution* [Darwinian]: study of learning and cooperation through dual heritage;
- *evolutionary linguistics* [Darwinian]: development of the elements of language;
- *visual communication* [Darwinian]: exploration of visual communication in primates;
- *emotion* [Darwinian]: individual-level study of emotional states.

Scale 3: small-scale social behavior (up to 2000 individuals).
Time frame: since speech, 70,000 years ago.
Disciplines:

- *historical linguistics*—language communities;
- *anthropology*—descent groups and networks, expanding through egalitarian and hierarchical structures;
- *cultural evolution* [Darwinian]: study of learning and cooperation through dual heritage;
- *emotions*—construction of emotions in individuals and groups;
- *institutional evolution* [Darwinian]—conscious formation and renewal of institutions by groups in collective intentionality, with communities averaging 150 and consolidating to larger groups during and after the Glacial Maximum. The Human System forms as a species-wide linkage of local networks.

Scale 4: large-scale social behavior (with agriculture).
Time frame: since agriculture, 10,000 years ago.
Disciplines:

- *historical linguistics*—language communities;
- *anthropology*—societies and states, achieving energy capture the expansion of collaborative and hierarchical organizations;
- *cultural evolution* [Darwinian]: study of learning and cooperation through dual heritage; "ultrasociety" as the extension of this logic beyond tribes to states;
- *institutional evolution* [Darwinian]—conscious formation and renewal of institutions by groups in collective intentionality, forming societies of about 2000 members, some of which grew significantly. The Human System deepens with the rise of interregional networks.

Scale 5: civilizational-scale social structure and behavior.
Time frame: since civilizations, 5000 years ago.
Disciplines:

- *civilization*—informal analysis in history and other fields, narrating and ranking early civilizations;
- *anthropology*—societies and states, expanding through hierarchy and energy capture;
- *world-systems*—societies, states, networks; rise and fall generated by initiatives from semiperipheral regions;
- *institutional evolution* [Darwinian]—conscious formation and renewal of institutions by groups in collective intentionality; connections through networks of growing size to the level of Human System.

Scale 6: global social structure and behavior.
Time frame: today.
Disciplines:

- *civilization*—informal narrative of large-scale polities and religions in recent centuries;
- *world*-systems—societies, states, networks; rise and fall generated by initiatives from semiperipheral regions;
- *institutional evolution* [Darwinian]—conscious formation and renewal of institutions by groups in collective intentionality, in groups of all sizes. Human System deepens to include global institutions and networks.

In a concluding commentary on this display of evolutionary research projects, I offer concise remarks on the experience of disciplines in the order of their founding. The early notions of *civilization* have changed little with time; aspects of this outlook have been adopted by other disciplines. *Biology*, in place by 1860, continues to develop techniques and theoretical nuance. *Anthropology*, analyzing descent and other groups, continues to broaden its analysis with steadily improving data: theory focuses mostly on expanding hierarchy and demographic scale; this analysis is extended to the level of large-scale social behavior. *Historical linguistics*, from the late eighteenth century, has improved techniques but a similar agenda. *Cultural evolution* has expanded its scale of interest to large-scale social behavior. *Evolutionary linguistics* (with Chomsky as founder), *studies of emotions*, and *visual communication* arose in the late twentieth century with individual and small-scale work. *World-Systems analysis*, from the 1980s, combined its core early modern analysis with a longer-term time frame, combining several social sciences, in work at large scales. *Institutional evolution*, only recently formulated, seeks to apply the conscious creation of institutions (including networks forming a Human System) from the level of small-scale societies to the global level. Of course, this simplified summary of research today leaves out many important details.

Finally, I offer a note of encouragement to the reader in her or his role as a researcher. How can a single researcher assemble knowledge from numerous disciplines? Researchers work best in groups but, even in groups, the individual retains the task of assembling and integrating knowledge in order to exchange it. Many researchers, if specialized in one field, have benefited from completing course work or minor fields in other disciplines. These will be, in effect, a researcher's first, second, or third academic language. In addition, there may exist additional fields in which one can become self-trained, up to a certain level of skill. Further, thanks to work by dedicated scholars, the basic principles and advances of many fields of study have been very well written up in articles and books, including those cited here. A final argument for taking on the study of so many disciplines at once is systemic: the Human System functions through interrelated processes, and studying a wide range of those processes through multiple disciplines should lead the researcher to encounter resonances and linkages in the system that might not be visible to study at a piecemeal level.

REFERENCES

Selected Reference Works

Baron, Robert S., and Norbert L. Kerr. *Group Process, Group Decision, Group Action*, 2nd ed. Philadelphia: Open University Press.
Bawden, David, and Lyn Robinson. *Introduction to Information Science*. Chicago, IL: Neal-Schuman Publishing, 2012.
Forsyth, Donelson R. *Group Dynamics*, 6th ed. Belmont, CA: Wadsworth Cengage Learning, 2014.
Hertler, Steven C., Aurelio José Figueredo, Mateo Peñaherrera-Aguirre, et al. *Foundations of Library and Information Science*, 4th ed.
Lucassen, Jan, Leo Lucassen, and Patrick Manning, eds. *Migration History in World History: Multidisciplinary Approaches*. Leiden: Brill, 2010.
McElreath, Richard, and Robert Boyd. *Mathematical Models of Social Evolution: A Guide for the Perplexed*. Chicago: University of Chicago Press, 2007.
Ruddiman, William F. *Earth's Climate, Past and Future*, 3rd ed. New York: W. H. Freeman, 2014.
Shryock, Andrew, and Daniel Lord Smail, eds. *Deep History: The Architecture of Past and Present*. Berkeley: University of California Press, 2011.

Websites Cited

American Chemical Society National Historic Chemical Landmarks. "Discovery of Radiocarbon Dating." http://www.acs.org/content/acs/en/education/whatischemistry/landmarks/radiocarbon-dating.html. Accessed May 26, 2018.
Arnett, Joe. "Coevolution and Pollination." *Washington Native Plant Society*, 2014. www.wnps.org.
Ask a Biologist. "Peppered Moths: Natural Selection Game." http://askabiologist.asu.edu/peppered-moths-game/kettlewell.

Bogumil, Elizabeth, and Christopher Chase-Dunn. "Settlement Networks and Sociocultural Evolution." IROWS Working Paper #127. http://irows.ucr.edu/papers/irows127/irows127.htm.
Caribbean Data Portal. http://www.caribbeaneconomics.org/about.
"Ice Core Basics." http://www.antarcticglaciers.org/glaciers-and-climate/ice-cores/ice-core-basics/.
Manning, Patrick. "Language Resources." www.cambridge.org/humanity.
Santa Fe Institute. "Evolution of Human Languages: An International Project on the Linguistic Prehistory of Humanity." http://ehl.santafe.edu.
Seshat. http://seshatdatabank.info.
Weart, Spencer. "The Discovery of Global Warming." http://www.aip.org/history/climate/index.htm.
World Inequality Database. http://WID.world.

Works Cited

Abbott, Patrick, J. Abe, J. Alcock, S. Alizon, et al. "Inclusive Fitness Theory and Eusociality." *Nature* 471, 7339 (2011): E1–4.
Adolphs, Ralph, and David J. Anderson. *The Neuroscience of Emotion: A New Synthesis*. Princeton: Princeton University Press, 2018.
Aiello, Leslie C., and R. I. M. Dunbar. "Neocortex Size, Group Size, and the Evolution of Language." *Current Anthropology* 34 (1993): 184–93.
Allen, Benjamin, Martin A. Nowak, and Edward O. Wilson. "Limitations of Inclusive Fitness." *PNAS* 110, 50 (2013): 20135–39.
Allen, Garland E. *Thomas Hunt Morgan: The Man and His Science*. Princeton: Princeton University Press, 1978.
Alston, Eric, Lee J. Alston, Bernardo Mueller, and Tomas Nonnenmacher. *Institutional and Organizational Analysis: Concepts and Applications*. Cambridge: Cambridge University Press, 2018.
Avery, O. T., C. M. MacLeod, and M. McCarty. "Studies on the Chemical Nature of the Substance Inducing Transformation of Pneumococcal Types: Induction of Transformation by a Desoxyribosenucleic Acid Fraction Isolated from Pneumococcus Type III." *Journal of Experimental Medicine* 79 (1944): 137–58.
Axelrod, Robert. *The Evolution of Cooperation*. New York: Basic Books, 1980.
Ayala, Francisco J. "Introduction." In Ayala and Dobzhansky, eds., *Studies in the Philosophy of Biology*, vii–xvi. Berkeley: University of California Press, 1974.
Ayala, Francisco J. "The Concept of Biological Progress." In Ayala and Dobzhansky, eds., *Studies in the Philosophy of Biology*, 339–55. Berkeley: University of California Press, 1974.
Ayala, Francisco J., and Theodosius Dobzhansky, eds. *Studies in the Philosophy of Biology: Reduction and Related Problems*. London: Macmillan, 1974.
Bacharach, Michael, Natalie Gold, and Robert Sugen, eds. *Beyond Individual Choice: Teams and Frames in Game Theory*. Princeton: Princeton University Press, 2006.
Bandura, Albert. *Social Learning Theory*. Morristown, NJ: General Learning Press, 1971.
Barrett, Lisa Feldman. *How Emotions Are Made: The Secret Life of the Brain*. Boston: Houghton Mifflin Harcourt, 2018.

Bateman, Richard, et al. "Speaking of Forked Tongues: The Feasibility of Reconciling Human Phylogeny and the History of Language [and Comments]." *Current Anthropology* 31, 1 (1990): 1–24.
Becker, Gary. "The Economic Way of Looking at Life." Nobel Prize Lecture. https://www.nobelprize.org/uploads/2018/06/becker-lecture.pdf.
Benedict, Ruth. *Race: Science and Politics.* New York: Modern Age Books, 1940.
Berger, A., and M. F. Loutre. "Insolation Values for the Climate ..." *Quaternary Science Reviews* 10 (1991): 297–317.
Berwick, Robert, and Noam Chomsky. *Why Only Us: Language and Evolution.* Cambridge, MA: MIT Press, 2016.
Bickerton, Derek. *Language and Species.* Chicago: University of Chicago Press, 1981.
Biglan, Anthony. "The Characteristics of Subject Matter in Different Academic Areas." *Journal of Applied Psychology* 57 (1973): 195–203.
Blackmore, Susan. *The Meme Machine.* Oxford: Oxford University Press, 1999.
Blau, Peter M. *Exchange & Power in Social Life.* New Brunswick, NJ: Transaction Publishers, 1986.
Boas, Franz. *Anthropology and Modern Life.* Westport, CT: Greenwood, 1984 [1928].
Boas, Franz. *The Mind of Primitive Man.* New York: Macmillan, 1938 [1911].
Bowler, Peter J. *Evolution.* Oakland: CA: University of California Press, 2009. First published in 1984 as *Evolution: The History of an Idea.*
Bowler, Peter J. *Evolution: The History of an Idea,* 3rd ed. Berkeley: University of California Press, 2003.
Boyd, Robert. *A Different Kind of Animal: How Culture Transformed Our Species.* Princeton: Princeton University Press, 2018.
Boyd, Robert, and Peter J. Richerson, "A Simple Dual Inheritance Model of the Human Evolutionary Process I: Basic Postulates and a Simple Model." Journal of the American Chemical Society 1 (1978): 127–54.
Boyd, Robert, and Peter J. Richerson, "A Simple Dual Inheritance Model of the Conflict Between Social and Biological Evolution." *Zygon, Journal of Religion and Science* 11 (1976): 129–31.
Boyd, Robert, and Peter J. Richerson. "Culture and the Evolution of Human Cooperation." *Philosophical Transactions of the Royal Society* 364 (2009): 3281–88.
Boyd, Robert, and Peter J. Richerson. *Culture and the Evolutionary Process.* Chicago: University of Chicago Press, 1985.
Boyd, Robert, and Peter J. Richerson. *The Origin and Evolution of Cultures.* New York: Oxford University Press, 2005.
Boyd, Robert, and Peter J. Richerson, "Why Culture Is Common But Cultural Evolution Is Rare." *Proceedings of the British Academy* 88 (1996): 73–93.
Braudel, Fernand, trans. Siân Reynolds. *The Mediterranean and the Mediterranean World in the Age of Philip II,* 2 vols. New York: Harper & Row, 1976.
Brooke, John L. *Climate Change and the Course of Global History: A Rough Journey.* Cambridge: Cambridge University Press, 2014.
Burke, James, and Robert Ornstein. *The Axemaker's Gift: A Double-Edged History of Human Culture.* New York: Grosser/Putnam, 1995.
Campbell, Bruce M. S. *The Great Transition: Climate, Disease and Society in the Late-Medieval World.* Cambridge: Cambridge University Press, 2016.
Campbell, Donald T. "'Downward Causation' in Hierarchically Organised Biological Systems." In Ayala and Dobzhansky, eds., *Studies in the Philosophy of Biology,* 179–86. Berkeley: University of California Press, 1974.

Campbell, Donald T. "Evolutionary Epistemology." In P. A. Schilpp, ed., *The Philosophy of Karl R. Popper*, 412–63. LaSalle, IL: Open Court, 1974.

Campbell, Donald T. "On the Conflicts Between Biological and Social Evolution and Between Psychology and Moral Tradition" (presidential address, American Psychological Association). *American Psychologist* (December 1975): 1103–26; reprinted as Donald T. Campbell, "On the Conflicts between Biological and Social Evolution and between Psychology and Moral Tradition." *Zygon* 11, 3 (1976): 167–208.

Campbell, Donald T. "The Two Distinct Routes Beyond Kin Selection to Ultrasociality: Implications for the Humanities and Social Sciences." In Diane L. Bridgeman, ed., *The Nature of Prosocial Development: Interdisciplinary Theories and Strategies.* New York: Academic Press, 1983.

Campbell, Donald T. "Unjustified Variation and Selective Retention in Scientific Discovery." In Ayala and Dobzhansky, eds., *Studies in the Philosophy of Biology*, 139–61. Berkeley: University of California Press, 1974.

Campbell, Donald T. "Variation and Selective Retention in Socio-Cultural Evolution." In Herbert R. Barringer, George I. Blandsten, and Raymond W. Mack, eds., *Social Change in Developing Areas*, 19–49. Cambridge, MA: Shenkman Publishing Company, 1965.

Campbell, Michael C., and Sarah A. Tishkoff. "The Evolution of Human Genetic and Phenotypic Variation in Africa." *Current Biology* 20 (2010): R166–R173.

Cann, Rebecca L., Mark Stoneking, and Allan C. Wilson. "Mitochondrial DNA and Human Evolution." *Nature* 325 (1987): 31–36.

Carneiro, Robert L. "Scale Analysis, Evolutionary Sequences, and the Rating of Cultures." In R. Naroll and R. Cohen, eds., *A Handbook of Method in Cultural Anthropology.* Garden City, NY: Doubleday, 1970.

Carneiro, Robert L. "The Transition from Quantity to Quality: A Neglected Causal Mechanism in Accounting for Social Evolution." *PNAS* 97 (2000): 12926–31.

Carneiro, Robert L. *Evolutionism and Cultural Anthropology: A Critical History.* Boulder: Westview Press, 2003.

Carrington, Peter J., John Scott, and Stanley Wasserman. *Models and Methods in Social Network Analysis.* Cambridge: Cambridge University Press, 2005.

Cavalli-Sforza, L. L. *Genes, Peoples, and Language.* New York: North Point Press, 2000.

Cavalli-Sforza, L. L., and M. W. Feldman. *Cultural Transmission and Evolution.: A Quantitative Approach.* Princeton: Princeton University Press, 1981.

Cavalli-Sforza, L. L., and M. W. Feldman. "Models for Cultural Inheritance. I. Group Mean and Within Group Variation." *Theoretical Population Biology* 4 (1973): 42–55.

Cavalli-Sforza, L. L., and M. W. Feldman. "Towards a Theory of Cultural Evolution." *Interdisciplinary Science Reviews* 3 (1978): 99–107.

Cavalli-Sforza, L. L., Paolo Menozzi, and Alberto Piazza. *History and Geography of Human Genes.* Princeton: Princeton University Press, 1994.

Cavalli-Sforza, Luigi Luca, Alberto Piazza, Paolo Menozzi, and Joanna Mountain. "Reconstruction of Human Evolution: Bringing Together Genetic, Archaeological, and Linguistic Data." *PNAS* 85, 16 (1988): 6002–6006.

Ceruzzi, Paul E. *Computing: A Concise History.* Cambridge, MA: MIT Press, 2012.

Charlson, Robert J., James E. Lovelock, Meinraj O. Andraea, and Stephen G. Warren. "Oceanic Phytoplankton, Atmospheric Sulphur, Cloud Albedo and Climate." *Nature* 326 (1987): 655–61.
Chase-Dunn, Christopher. *The Wintu and Their Neighbors: A Very Small World-System in Northern California*. Tucson: University of Arizona Press, 1998.
Chase-Dunn, Christopher, and Bruce Lerro. *Social Change: Globalization from the Stone Age to the Present*. Boulder: Paradigm Publishers, 2014.
Chase-Dunn, Christopher, and Thomas D. Hall. *Rise and Demise: Comparing World-Systems*. Boulder, CO: Westview Press, 1997.
Childe, V. Gordon. *Social Evolution*. London: Watts & Co., 1951.
Chomsky, Noam. *Syntactic Structures*. 's-Gravenhage: Mouton, 1957.
Chomsky, Noam. *The Minimalist Program*. Cambridge, MA: MIT Press, 1995.
Christian, David. *Maps of Time: An Introduction to Big History*. Berkeley: University of California Press, 2003.
Christian, David. *Origin Story: A Big History of Everything*. New York: Little, Brown, 2018.
Christopher Chase-Dunn, Hiroko Inoue, Eugene Anderson, and David Wilkinson. "The SETPOL Framework: Settlements and Polities in World-Systems." IROWS Working Paper #114. irows.ur.edu/papers/irows114/irows114.htm.
Coleman, James S. "Social Theory, Social Research, and a Theory of Action." *American Journal of Sociology* 91 (1986): 1309–35.
Cox, F. E. G. "History of Human Parasitology." *Clinical Microbiology Reviews* (2002): 595–612.
Creanza, Nicole, Oren Kolodny, and Marcus W. Feldman, "Cultural Evolutionary Theory: How Culture Evolves and Why It Matters." *PNAS* 114, 30 (2017): 7782–89.
Croft, William. *Joseph Harold Greenberg, 1915–2001: A Biographical Memoir*. Washington, DC: National Academy of Sciences, 2007.
Crosby, Alfred W. *The Columbian Exchange: Biological Consequences of 1492*. Westport, CT: Greenwood Publishers, 1972.
Darrigol, Olivier. *Atoms, Mechanics, and Probability: Ludwig Boltzmann's Statistico-Mechanical Writings—An Exegesis*. Oxford: Oxford University Press, 2018.
Darwin, Charles. *The Descent of Man and Selection in Relation to Sex*, 2nd ed. New York: A. L. Burt, 1874.
Darwin, Charles. *The Expression of Emotions in Man and Animals*, with an introduction, afterword, and commentaries by Paul Ekman, 3rd ed. Oxford: Oxford University Press, 1998 [1872].
Darwin, Charles. *The Origin of Species by Means of Natural Selection*. London: John Murray, 1859.
Dawkins, Richard. *The Selfish Gene*. Oxford: Oxford University Press, 1979.
de Zwart, Pim, and Jan Luiten van Zanden. *The Origins of Globalization: World Trade in the Making of the Global Economy, 1500–1800*. Cambridge: Cambridge University Press, 2018.
Degler, Carl N. *In Search of Human Nature: The Decline and Revival of Darwinism in American Social Thought*. New York: Oxford University Press, 1991.
Delbrück, Max. *Mind from Matter? An Essay in Evolutionary Epistemology*. Oxford: Blackwell, 1986.

Diamond, Jared. *Guns, Germs, and Steel: The Fates of Human Societies.* New York: Norton, 1997.
Dillehay Tom D., et al. "New Archaeological Evidence for an Early Human Presence at Monte Verde, Chile." *PLoS One* (November 18, 2015): 1–5.
Dilthey, Wilhelm, trans. Ramon J. Betanzos. *Introduction to the Human Sciences.* Detroit: Wayne State University Press, 1988 [1883].
Dingle, Hugh. *Migration: The Biology of Life on the Move.* Oxford: Oxford University Press, 1996.
Dobzhansky, Theodosius. "Chance and Creativity in Evolution." In Ayala and Dobzhansky, eds., *Studies in the Philosophy of Biology,* 307–38. Berkeley: University of California Press, 1974.
Dobzhansky, Theodosius. *Genetics and the Origin of Species,* introduced by Stephen Jay Gould. New York: Columbia University Press 1982 [1937].
Drwenski, Matt. "Scales of Inequality: Strategies for Researching Global Disparities from 1750 to the Present." MA thesis, University of Pittsburgh, 2015.
Dunbar, Robin. *Grooming, Gossip and the Evolution of Language.* London: Faber and Faber, 1996.
Dunbar, Robin I. M., and Richard Sosis. "Optimising Human Community Sizes." *Evolution and Human Behavior* 39 (2018): 106–11.
Durham, William H. *Coevolution: Genes, Culture, and Human Diversity.* Stanford: Stanford University Press, 1991.
Durkheim, Émile. *De la division du travail social.* Paris: Alcan, 1922 [1893].
Ehret, Christopher. *A Historical-Comparative Reconstruction of Nilo-Saharan.* Köln: R. Köppe Verlag, c2001.
Ehret, Christopher. *History and the Testimony of Language.* Berkeley: University of California Press, 2011.
Ehret, Christopher. "Nostratic—or Proto-Human?" In Colin Renfrew and Daniel Nettle, eds., *Nostratic: Examining a Linguistic Macrofamily,* 93–122. Cambridge: The McDonald Institute for Archaeological Research, 1998.
Ehret, Christopher. *Reconstructing Proto-Afroasiatic (Proto-Afrasian): Vowels, Tone, Consonants, and Vocabulary.* Berkeley: University of California Press, c1995.
Ehrlich, Paul R., and Peter H. Raven. "Butterflies and Plants: A Study in Coevolution." *Evolution* 18 (1964): 586–608.
Elias, Norbert, trans. Stephen Mennell, and Grace Morrissey. *What Is Sociology?* New York: Columbia University Press, 1978.
Engels, Frederick. "The Origin of the Family, Private Property, and the State." In Karl Marx and Frederick Engels, eds., *Selected Works,* 468–593. New York: International Publishers, 1969.
Eustace, Nicole, Eugenia Lean, Julie Livingston, Jan Plamper, William M. Reddy, and Barbara H. Rosenwein. "*AHR* Conversation: The Historical Study of Emotions." *American Historical Review* 117 (2012): 1487–531.
Feldman, M. W., and L. L. Cavalli-Sforza. "Cultural Transmission." In M. Pagel, ed., *Encyclopedia of Evolution,* 222–26. New York: Oxford University Press, 2002.
Feldman, M. W., L. L. Cavalli-Sforza, and L. A. Zhivotovsky. "On the Complexity of Cultural Transmission and Evolution." In G. Cowan, D. Pines, and D. Meltzer, eds., *Complexity: Metaphors, Models, and Reality,* 47–62. Boston: Addison Wesley, 1994.
Felsenfeld, Gary. "A Brief History of Epigenetics." *Cold Spring Harbor Perspectives in Biology* 6 (2014): PMC3941222.

Ferguson, Adam. *Essay on the History of Civil Society*. Edinburgh, 1768.
Fernández-Armesto, Felipe. *A Foot in the River: Why Our Lives Change—And the Limits of Evolution*. Oxford: Oxford University Press, 2015.
Ferriere, Regis, and Richard E. Michod. "Inclusive Fitness in Evolution." *Nature* 477 (2011): E6–E8.
Finlay, Barbara L. "Evolution, Development, and Emerging Universals." In Morten H. Christiansen, Chris Collins, and Shimon Edelman, eds., *Language Universals*, 261–65. Oxford: Oxford University Press, 2009.
Fisher, Ronald A. *The Genetical Theory of Natural Selection*. Oxford: Clarendon Press, 1930.
Fisher, Ronald A. *The Genetical Theory of Natural Selection*, 2nd ed. New York: Dover, 1958.
Fitch, W. Tecumseh. *The Evolution of Language*. Cambridge: Cambridge University Press, 2010.
Flannery, Kent, and Joyce Marcus. *The Creation of Inequality: How Our Prehistoric Ancestors Set the Stage for Monarchy, Slavery, and Empire*. Cambridge, MA: Harvard University Press, 2012.
Forrester, Jay W. *World Dynamics*. Cambridge, MA: Wright-Allen Press, 1971.
Frank, Andre Gunder, and Barry K. Gills. "The Five Thousand Year World System: An Interdisciplinary Introduction." *Humboldt Journal of Social Relations* 18, 2 (1993): 1–79.
Fressoz, Jean-Baptiste, and Christophe Bonneuil. "Growth Unlimited: The Idea of Infinite Growth from Fossil Capitalism to Green Capitalism." In Iris Borowy and Matthias Schmelzer, eds., *History of the Future of Economic Growth: Historical Roots of Current Debates on Sustainable Growth*, 52–68. London: Routledge, 2017.
Gardner, A., S. A. West, and G. Wild. "The Genetical Theory of Kin Selection." *Journal of Evolutionary Biology* 24, 5 (2011): 1020–43.
Gaur, Albertine. *A History of Writing*, rev. ed. London: The British Library, 1992.
Gillette, Aaron. *Eugenics and the Nature-Nurture Debate in the Twentieth Century*. New York: Palgrave Macmillan, 2007.
Gills, Barry K., and Andre Gunder Frank, eds. *The World System: Five Hundred Years or Five Thousand?* London: Routledge, 1994.
Gobineau, Arthur, comte de, trans. Adrian Collins. *The Inequality of Human Races*. Los Angeles: Noontide Press, 1966.
Gogol, Nikolai, trans. Alexander Tulloch. *Arabesques*. Ann Arbor: Ardis, 1982 [1835].
Goodrich, Samuel G. *Peter Parley's Universal History on the Basis of Geography, by Samuel Griswold*. 1859.
Gould, Stephen Jay. *Ontogeny and Phylogeny*. Cambridge, MA: Belknap Press of Harvard University Press, 1977.
Gould, Stephen Jay. *Punctuated Equilibrium*. Cambridge, MA: Belknap Press of Harvard University Press, 2007.
Gould, Stephen Jay. *The Structure of Evolutionary Theory*. Cambridge, MA: Belknap Press of Harvard University Press, 2002.
Gran, Peter. *The Rise of the Rich: A New View of Modern World History*. Syracuse: Syracuse University Press, 2009.
Greenberg, Joseph H. *Indo-European and Its Nearest Neighbors: The Eurasiatic Language Family*, 2 vols. Stanford: Stanford University Press, 2000.
Greenberg, Joseph H. *Language in the Americas*. Stanford: Stanford University Press, 1987.

Greenberg, Joseph H. *Languages of Africa*. Bloomington, IN: Indiana University Press, 1963.
Greenberg, Joseph H. "Linguistic Evidence Regarding Bantu Origins." *Journal of African History* 13 (1972): 189–216.
Greenberg, Joseph H., ed. *Universals of Language*, 2nd ed. Cambridge, MA: MIT Press, 1965.
Greenberg, Joseph H. "The Indo-Pacific Hypothesis (1971)." In Greenberg (ed. Croft), *Genetic Linguistics*, chapter 12.
Greenberg, Joseph H., Charles A. Ferguson, and Edith A. Moravcsik, eds. *Universals of Human Language*, 4 vols. Stanford: Stanford University Press, 1978.
Greenberg, Joseph H., Christy G. Turner II, Stephen L. Zegura, Lyle Campbell, James A. Fox, W. S. Laughlin, Emöke J. E. Szathmary, Kenneth M. Weiss, and Ellen Woolford. "The Settlement of the Americas: A Comparison of the Linguistic, Dental, and Genetic Evidence [and Comments and Reply]." *Current Anthropology* 27 (1986): 477–97.
Greenberg, Joseph H., edited and introduced by William Croft. *Genetic Linguistics: Essays on Theory and Method*. Oxford: Oxford University Press, 2018.
Halbwachs, Maurice, ed. and trans. Lewis A. Coser. *On Collective Memory*. Chicago: University of Chicago Press, 1992 [1950].
Hallpike, C. R. *Foundations of Primitive Thought*. Oxford: Clarendon Press, 1979.
Hallpike, C. R. *How We Got Here: From Bows and Arrows to the Space Age*. Central Milton Keynes, UK: Authorhouse, 2008.
Hallpike, C. R. *The Principles of Social Evolution*. Oxford: Clarendon Press, 1986.
Hamilton, W. D. "The Genetical Evolution of Social Behaviour, I." *Journal of Theoretical Biology* 7 (1964): 1–16.
Hamilton, W. D. "The Genetical Evolution of Social Behaviour, II." *Journal of Theoretical Biology* 7 (1964): 17–52.
Hancock, David. "Commerce and Conversation in the Eighteenth-Century Atlantic: The Invention of Madeira Wine." *Journal of Interdisciplinary History* 29 (1998): 197–219.
Hansen, James. *Storms of My Grandchildren: The Truth about the Coming Climate Catastrophe and Our Last Chance to Save Humanity*. London: Bloomsbury, 2009.
Harari, Yuval Noah. *Sapiens: A Brief History of Humankind*. New York: Harper, 2015.
Harms, William F. *Information and Meaning in Evolutionary Processes*. Cambridge: Cambridge University Press, 2004.
Haspelmath, Martin. "Preface." In Joseph H. Greenberg, ed., *Language Universals, with Special Reference to Feature Hierarchies*, vii. Berlin: Mouton de Gruyter, 2005 [1966].
Hayek, Friedrich A. *Studies in Philosophy, Politics and Economics*. Chicago: University of Chicago Press, 1967.
Hayek, Friedrich A. *The Counter-Revolution of Science*. New York: Free Press, 1955.
Heath, Joseph. "Methodological Individualism." In Edward N. Zalta, ed., *The Stanford Encyclopedia of Philosophy*, Summer 2013 Edition. https://plato.stanford.edu/archives/sum2013/entries/methodological-individualism/.
Hennig, Willi, trans. D. Davis and R. Zangerl. *Phylogenetic Systematics*. Urbana: University of Illinois Press, 1966 [1950].
Henrich, Joseph. "Cooperation, Punishment, and the Evolution of Human Institutions." *Science* 312 (2006): 60–61.

Henrich, Joseph. *The Secret of our Success: How Culture Is Driving Human Evolution, Domesticating Our Species, and Making Us Smarter.* Princeton: Princeton University Press, 2016.
Henrich, Joseph, and R. McElreath. "The Evolution of Cultural Evolution" *Evolutionary Anthropology* 12 (2003): 123–35.
Henrich, Joseph, Robert Boyd, and Peter J. Richerson, "Five Misunderstandings About Cultural Evolution." *Human Nature* 19 (2008): 119–37.
Hinshelwood, Cyril N., and Linus Pauling. "Amedeo Avogadro." *Science* 124 (1956): 708–13.
Hirbo, Jibril, A. Ranciaro, and S. A. Tishkoff. "Population Structure and Migration in Africa: Correlations Between Archaeological, Linguistic and Genetic Data." In *Causes and Consequences of Human Migration: An Evolutionary Perspective*, C.B.C. (2009): 135–71. https://doi.org/10.1017/cbo9781139003308.011.
Hoerder, Dirk. *Cultures in Contact: World Migrations in the Second Millennium.* Durham, NC: Duke University Press, 2002.
Hölldobbler, Bert, and Edward O. Wilson. *The Ants.* Cambridge, MA: Belknap Press of Harvard University Press, 1990.
Holmes, Arthur. *The Age of the Earth.* London: Harper Brothers, 1913.
Hornborg, Alf, and Carole L. Crumley, eds. *The World System and the Earth System: Global Socioenvironmental Change and Sustainability since the Neolithic.* New York: Routledge, 2016 [2006].
Houkes, Wybo. "Population Thinking and Natural Selection in Dual-Inheritance Theory." Biology and Philosophy 27 (2012): 401–17.
Huxley, Julian. *Evolution: The Modern Synthesis.* New York: Harper, 1942.
Ingold, Tim. *Evolution and Social Life.* Cambridge: Cambridge University Press, 1986.
Ingold Tim, and Gisli Paisson, eds., *Biosocial Becomings: Integrating Social and Biological Anthropology.* Cambridge: Cambridge University Press, 2013.
Jablonski, Nina "The Evolution of Human Skin and Skin Color." *Annual Review of Anthropology* 33 (2004): 585–623.
Jablonski, Nina G., and P. G. Maré, eds. *The Effects of Race.* Stellenbosch: African Sun Media, 2018.
Jacob, François. "Evolution and Tinkering." *Science* 196 (1977): 1161–66.
Johannsen, Wilhelm. "The Genotype Conception of Heredity." *The American Naturalist* 45 (1911): 129–59.
Kawai, Kaori, ed. *Groups: The Evolution of Human Sociality*, trans. Minako Sato. Kyoto: Kyoto University Press, 2013 [2009].
Kawai, Kaori, ed. *Institutions: The Evolution of Human Sociality*, trans. Minako Sato. Kyoto: Kyoto University Press, 2017 [2013].
Kellner, L. *Alexander von Humboldt.* London: Oxford University Press, 1963.
Kiel, L. Douglas, and Euel Elliott, eds. *Chaos Theory in the Social Sciences: Foundations and Applications.* Ann Arbor: University of Michigan Press, 1997.
Kolbert, Elizabeth. *The Sixth Extinction: An Unnatural History.* New York: Henry Holt, 2014.
Kroeber, Alfred L. *The Nature of Culture.* Chicago: University of Chicago Press, 1952.

Kroeber, Alfred L., and Clyde Kluckhohn, with the assistance of Wayne Unterciner and Appendices by Alfred G. Meyer. *Culture. A Critical Review of Concepts and Definitions* (Papers of the Peabody Museum of American Archaeology and Ethnology, Harvard University, Vol. XLVII-No. 1) Cambridge, MA: Published by the Museum, 1952.

Kuper, Adam. *The Invention of Primitive Society: Transformations of an Illusion.* London: Routledge, 1988.

Laland, Kevin N. "Cultural Evolution." In Mark Pagel, ed., *Encyclopedia of Evolution*, 220. Oxford: Oxford University Press, 2002.

Laland, Kevin N. *Darwin's Unfinished Symphony: How Culture Made the Human Mind.* Princeton: Princeton University Press, 2017.

Latchman, David. *Gene Regulation: A Eukaryotic Perspective*, 4th ed. Cheltenham, UK: Nelson Thornes, 2002.

Leakey, L. S. B., P. V. Tobias, and J. R. Napier. "A New Species of the Genus *Homo* from Olduvai Gorge." *Nature* 202 (1964): 7–9.

Leakey, Richard, and Roger Lewin. *The Sixth Extinction: Patterns of Life and the Future of Humankind.* New York: Doubleday, 1995.

Lederberg, Joshua, and Norton D. Zinder. "Concentration of Biochemical Mutants of Bacteria with Penicillin." *Journal of the American Chemical Society* 70 (1948): 4267.

Lewens, Tim. *Cultural Evolution: Conceptual Challenges.* Oxford: Oxford University Press, 2015.

Lewin, K. "Defining the 'Field at a Given Time.'" *Psychological Review* 50 (1943): 292–310.

Lewin, Kurt. "Frontiers in Group Dynamics II: Channels of Group Life; Social Planning and Action Research." *Human Relations* 1 (1947): 143–53.

Lewin, Kurt. "Frontiers in Group Dynamics: Concept, Method and Reality in Social Science; Social Equilibria and Social Change." *Human Relations* 1 (1947): 5, 9–10.

Lewis, Herbert S. "Boas, Darwin, Science, and Anthropology." *Current Anthropology* 42 (2001): 381–406.

Lewis, Herbert S. "The Individual and Individuality in Franz Boas's Anthropology and Philosophy." In Regna Darnell, Michelle Hamilton, Robert L. A. Hancock, and Joshua Smith, eds., *The Franz Boas Papers*, Vol. 1, 19–41. Lincoln, NB: University of Nebraska Press, 2015.

Lewontin, R. C. "Interview of R. C. Lewontin." In Rama S. Singh, Costas B. Krimbas, Diane B. Paul, and John Beatty, eds., *Thinking about Evolution: Historical, Philosophical and Political Perspectives*, Vol. 2, 22–61. Cambridge: Cambridge University Press, 2001.

Lewontin, R. C., Steven Rose, and Leon Kamin. *Not in our Genes: Biology, Ideology, and Human Nature.* New York: Pantheon, 1984.

Lieberman, Philip. "The Evolution of Human Speech: Its Anatomical and Neural Bases." *Current Anthropology* 48, 1 (2007): 39–66.

Lindert, Peter H., and Jeffrey G. Williamson. "Globalization and Inequality: A Long History." World Bank Annual Bank Conference on Development Economics—Europe, 2001.

Lipson, Mark, Isabelle Ribot, David Reich, et al. "Ancient West African Foragers in the Context of African Population History." *Nature* 577 (2020): 665–70.

Lopreato, Joseph. *Human Nature and Biocultural Evolution.* Boston: Allen & Unwin, 1984.
Lotka, Alfred J. *Elements of Physical Biology.* Baltimore: Williams & Wilkins, 1925.
Lovelock, James. *Gaia, A New Look at Life on Earth.* Oxford: Oxford University Press, 1979.
Lovelock, James. *The Revenge of Gaia: Earth's Climate Crisis and the Fate of Humanity.* New York: Basic Books, 2007.
Lovelock, James, and Sidney Epton. "The Quest for Gaia." *New Scientist* (February 8, 1975): 304.
Lucassen, Leo. "Working Together: New Directions in Global Labour History." *Journal of Global History* 11 (2016): 66–87.
Lumsden, Charles J., and Edward O. Wilson. *Genes, Mind, and Culture: The Coevolutionary Process.* Cambridge, MA: Harvard University Press, 1981.
Lyell, Charles. *Principles of Geology*, 3 vols. London: J. Murray, 1830–33.
Lynn, Richard. *Eugenics: A Reassessment.* Westport, CT: Praeger, 2001.
Majerus, Michael E. N. 2008. "Industrial Melanism in the Peppered Moth, *Biston betularia*: An Excellent Teaching Example of Darwinian Evolution in Action." *Evolution: Education and Outreach* 2, 1 (2008): 63–67.
Manning, Patrick. *A History of Humanity: The Evolution of the Human System.* Cambridge: Cambridge University Press, 2020.
Manning, Patrick. *Big Data in History.* London: Palgrave Macmillan, 2013.
Manning, Patrick. "Cross-Community Migration: A Distinctive Human Pattern." *Social Evolution and History* 5, 2 (2006): 24–54.
Manning, Patrick. "Epistemology." In Jerry H. Bentley, ed., *Oxford History Handbook: World History*, 105–21. New York: Oxford University Press, 2011.
Manning, Patrick. "Homo Sapiens Populates the Earth: A Provisional Synthesis, Privileging Linguistic Data." *Journal of World History* 17 (2006): 115–58.
Manning, Patrick. "Inequality: Historical and Disciplinary Approaches." *American Historical Review* 122 (2017): 1–22.
Manning, Patrick. *Navigating World History: Historians Create a Global Past.* New York: Palgrave Macmillan, 2003.
Manning, Patrick. "The Life-Sciences, 1900–2000: Analysis and Social Welfare from Mendel and Koch to Biotech and Conservation." *Asian Review of World Histories* 6 (2018): 185–208.
Manning, Patrick, and Abigail Owen, eds. *Knowledge in Translation: Global Patterns of Scientific Exchange, 1000–1800 CE.* Pittsburgh: University of Pittsburgh Press, 2018.
Manning, Patrick, and Aubrey Hillman. "Climate as a Factor in Migration and Social Change, 200,000 to 5000 Years Ago." American Historical Association annual meeting, New Orleans, 5 January 2013.
Manning, Patrick, and Daniel Roods, eds. *Global Scientific Practice in an Age of Revolutions, 1750–1850.* Pittsburgh: University of Pittsburgh Press, 2016.
Manning, Patrick, and Mat Savelli, eds. *Global Transformation in the Life Sciences, 1945–1980.* Pittsburgh: University of Pittsburgh Press, 2018.
Manning, Patrick, and Sanjana Ravi. "Cross-Disciplinary Theory in Construction of a World-Historical Archive." *Journal of World-Historical Information* 1 (2013): 15–39.

Manning, Patrick, with Tiffany Trimmer. *Migration in World History*, 3rd ed. London: Routledge, 2020.
Marks, Jonathan. *Human Biodiversity: Genes, Race, and History*. New York: Aldine De Gruyter, 1995.
Marx, Karl, ed. and trans. David McLellan. *The Grundrisse*. New York: Harper & Row, 1971.
Marx, Karl. *Capital*, 3 vols. Moscow: Progress Publishers, 1971.
Matt, Susan J., and Peter N. Stearns, eds. "Doing Emotions History." *American Historical Review* 120 (2014): 187–88.
Maxwell, James Clerk. "A Dynamical Theory of the Electromagnetic Field." *Philosophical Transactions of the Royal Society of London* 155 (1865): 512. https://doi.org/10.1098/rstl.1865.0008.
Maynard Smith, John, and Eörs Szathmáry. *The Major Transitions in Evolution*. Oxford: Oxford University Press, 1995.
Maynard Smith, John. "Game Theory and The Evolution of Fighting." In *On Evolution*. Edinburgh: Edinburgh University Press, 1972.
Maynard Smith, John. "Group Selection and Kin Selection." *Nature* 201 (1964): 1145–47.
Maynard Smith, John. *Evolution and the Theory of Games*. Cambridge: Cambridge University Press, 1982.
Maynard Smith, John, and George R. Price. "The Logic of Animal Conflict." *Nature* 246, 5427 (1973): 15–18.
Mayr, Ernst. *Evolution and the Diversity of Life*. Cambridge, MA: Harvard University Press, 1975.
McBrearty, Sally, and Alison Brooks. "The Revolution That Wasn't: A New Interpretation of the Origin of Modern Human Behavior." *Journal of Human Evolution* 39 (2000): 453–563.
McElreath, Richard. "A Long-form Research Program in Human Behavior, Ecology and Culture," 2017.
McEvedy, Colin, and Richard Jones. *Atlas of World Population History*. Harmondsworth, UK: Penguin, 1978.
McLean, Paul. *Culture in Networks*. Cambridge: Polity Press, 2017.
McNeill, J. R. *Something New Under the Sun: An Environmental History of the Twentieth-Century World*. New York: Norton, 2000.
McNeill, J. R., and William H. McNeill. *The Human Web: A Bird's-Eye View of Human History*. New York: Norton, 2003.
McNeill, William H. *Keeping Together in Time: Dance and Drill in Human History*. Cambridge, MA: Harvard University Press, 1995.
McNeill, William H. *The Rise of the West: A History of the Human Community*. Chicago: University of Chicago Press, 1963.
Meadows, Donnella H., Dennis L. Meadows, Jørgen Randers, and William W. Behrens II. *The Limits to Growth: A Report for the Club of Rome's Project on the Predicament of Mankind*. New York: Universe Books, 1972.
Meadows, Donnella H., Jorgen Randers, and Dennis Meadows. *The Limits to Growth: The 30-Year Update*. White River Junction, VT: Chelsea Green Publishing, 2004.
Mesoudi, Alex, and Kenichi Aoki, eds. *Learning Strategies and Cultural Evolution in the Palaeolithic*. Tokyo and Heidelberg: Springer, 2015.
Mesoudi, Alex. *Cultural Evolution: How Darwinian Theory Can Explain Human Culture and Synthesize the Social Sciences*. Chicago: University of Chicago Press, 2011.

Mih, Walter. *The Fascinating Life and Theory of Einstein.* Bloomington, IN: AuthorHouse, 2004.
Milankovitch, Milutin. "Mathematische Klimalehre und Astronomische Theorie der Klimaschwankungen," in *Handbuch der Klimatologie,* 5 vols., Band 1, Allgemeine Klimalehre. Teil A. Mathematische Klimalehre und astronomische.
Milanovic, Branko. *Capitalism, Alone: The Future of the System That Rules the World.* Cambridge, MA: Belknap Press of Harvard University Press, 2019.
Mill, John Stuart. *On Liberty.* London: John Parker and Son, 1859.
Miller, James G. *Living Systems.* New York: McGraw-Hill, 1978.
Miller, John H., and Scott E. Page, *Complex Adaptive Systems: An Introduction to Computational Models of Social Life.* Princeton: Princeton University Press, 2007.
Minamizuka, Shingo, ed. *World History Teaching in Asia.* Great Barrington, MA: Berkshire Publishing, 2019.
Mithen, Steven. *After the Ice: A Global Human History, 20,000–5000 BC.* Cambridge, MA: Harvard University Press, 2003.
Moore, Jason W. *Capitalism in the Web of Life: Ecology and the Accumulation of Capital.* London: Verso, 2014.
Morange, Michel. *A History of Molecular Biology,* trans. Matthew Cobb. Cambridge, MA: Harvard University Press, 1998.
Morgan, Lewis Henry. *Ancient Society; Or, Researches in the Lines of Human Progress from Savagery, Through Barbarism to Civilization.* London: Macmillan, 1877.
Morgenthau, Hans Joachim. *Politics Among Nations: The Struggle for Power and Peace.* New York: Alfred A. Knopf, 1948.
Morris, Ian. *The Measure of Civilization: How Social Development Decides the Fate of Nations.* Princeton, NJ: Princeton University Press, 2013.
Morris, Ian. *Why the West Rules—For Now: The Patterns of History, and What They Reveal About the Future.* New York: Farar, Straus, and Giroux, 2010.
Murdock, George Peter. *Africa Its Peoples and their Culture History.* New York: McGraw Hill: 1959.
Nagel, Ernest. *The Structure of Science.* New York: Harcourt, Brace and World, 1961.
Nash, John. "Non-cooperative Games." *Annals of Mathematics* 54 (1951): 286–95.
Naumann, Katja. *Laboratorien der Weltgeschichtsschreibung: Lehre und Forschung an den Universitäten Chicago, Columbia und Harvard 1918 bis 1968.* Göttingen: Vandenhoeck & Ruprecht, 2019.
Nirenberg, Marshall, and J. Heinrich Matthaei, "The Dependence of Cell-Free Protein Synthesis in E. Coli Upon Naturally Occurring or Synthetic Polyribonucleotides." *PNAS* 47, 10 (1961): 1588–602.
North, Douglass C., John Joseph Wallis, Steven B. Webb, and Barry R. Weingast. "Limited Access Orders: An Introduction to the Conceptual Framework." In Douglass C. North, John Joseph Wallis, Steven B. Webb, and Barry R. Weingast, eds., *In the Shadow of Violence: Politics, Economics, and the Problems of Development,* 1–23 New York: Cambridge University Press, 2013.
North, Douglass C., John Joseph Wallis, Steven B. Webb, and Barry R. Weingast. *Violence and Social Orders: A Conceptual Framework for Interpreting Recorded History.* New York: Cambridge University Press, 2009.
Novartis Foundation Symposium. *The Limits of Reductionism in Biology.* New York: Wiley, 1998.
Nowak, Martin A., Corina E. Tarnita, and Edward O. Wilson. "The Evolution of Eusociality." *Nature* 466 (2010): 1057–62.

Olson, Mancur. *The Logic of Collective Action: Public Goods and the Theory of Groups.* Cambridge, MA: Harvard University Press, 1965.

O'Rourke, Kevin H., and Jeffrey G. Williamson. *Globalization and History: The Evolution of a Nineteenth-Century Atlantic Economy.* Cambridge, MA: MIT Press, 1999.

Östlung, Johan, trans. Lena Olsson. *Humboldt and the Modern German University.* Lund: Lund University Press, 2016.

Pareto, Vilfredo, trans. Ann S. Schwier. *Manual of Political Economy.* New York: Augustus M. Kelley, 1971 (First published 1906; translated from French edition of 1927).

Pareto, Vilfredo. *The Rise and Fall of Elites: An Application of Theoretical Sociology*, introduced by Hans L. Zetterberg. New Brunswick: Transaction Publishers, 1991.

Pareto, Vilfredo. *Trattato di Sociologia Generale* (Florence: G. Barbèra, 1916) [English translation The Mind and Society, trans. Andrew Bongiorno and Arthur Livingston (New York: Harcourt, Brace, 1935)].

Pauling, Linus, and E. Bright Wilson, Jr. *Introduction to Quantum Mechanics, with Applications to Chemistry.* New York: McGraw-Hill, 1935.

Pearl, Judea, and Dana Mackenzie. *The Book of Why: The New Science of Cause and Effect.* New York: Basic Books, 2018.

Pinker, Steven, and Paul Bloom. "Natural Language and Natural Selection." *Behavioral and Brain Sciences* 13 (1990): 707–27.

Pinker, Steven. *The Better Angels of our Nature: Why Violence Has Declined.* New York: Viking, 2011.

Pinker, Steven. *The Blank Slate: The Modern Denial of Human Nature.* New York: Viking, 2002.

Pinker, Steven. *The Language Instinct: How the Mind Creates Language.* New York: Harper, 2007.

Pomeranz, Kenneth. *The Great Divergence: China, Europe, and the Making of the Modern World Economy.* Princeton: Princeton University Press, 2000.

Pritchard, Jonathan K., Joseph K. Pickrell, and Graham Coop. "The Genetics of Human Adaptation: Hard Sweeps, Soft Sweeps, and Polygenic Adaptation." *Current Biology* 20 (2010): R206–R215.

Ratcliff, Jessica. "The Great Data Divergence: Global History of Science Within Global Economic History." In *Manning and Rood, Global Scientific Practice*, 237–54.

Reich, David. *Who We Are and How We Got Here.* New York: Pantheon, 2018.

Richardson, Ken. *Genes, Brains, and Human Potential: The Science and Ideology of Intelligence.* New York: Columbia University Press, 2019.

Richerson, Peter J. "Recent Critiques of Dual Inheritance Theory." *Evolutionary Studies in Imaginative Culture* 1 (2017): 203–11.

Richerson, Peter J., and Morten Christiansen, eds. *Cultural Evolution: Society, Technology, Language, and Religion.* Cambridge, MA: MIT Press, 2013.

Richerson, Peter J., and Robert Boyd. "The Darwinian Theory of Human Cultural Evolution and Gene–Culture Coevolution." In Michael A. Bell, Douglas J. Futuyma, Walter F. Eanes, and Jeffrey S. Levinton, eds., *Evolution Since Darwin: The First 150 Years*, 561–88. Sunderland, MA: Sinauer Associates, Inc., 2010.

Richerson, Peter J., and Robert Boyd. "The Evolution of Human Ultra-Sociality." In Irenäus Eibl-Eibisfeldt and F. Salter, eds., *Indoctrinability, Ideology, and Warfare*, 71–95. New York: Berghahn Books, 1998.

Richerson, Peter J., and Robert Boyd. *Not by Genes Alone: How Culture Transformed Human Evolution*. Chicago: University of Chicago Press, 2005.
Rosenwein, Barbara H., and Riccardo Cristiani. *What Is the History of Emotions?* Cambridge: Polity Press, 2018.
Rousseau, Jérôme. *Rethinking Social Evolution: The Perspective from Middle-Range Societies*. Montreal: McGill-Queen's University Press, 2006.
Ruse, Michael. "Charles Darwin and Group Selection." *Annals of Science* 37 (1980): 615–30.
Russell, Edmund. *Evolutionary History: Uniting History and biology to Understand Life on Earth*. New York: Cambridge University Press, 2011.
Sanderson, Stephen. *Evolutionism and Its Critics: Deconstructing and Reconstructing an Evolutionary Interpretation of Human Society*. Boulder, CO: Paradigm Publishers, 2007.
Saussure, Ferdinand de, trans. Wade Baskin. *Course in General Linguistics*. New York: McGraw Hill, 1966 [1916].
Scheidel, Walter. *The Great Leveler: Violence and the History of Inequality from the Stone Age to the Twenty-first Century*. Princeton: Princeton University Press, 2017.
Schweikard, David P., and Hans Bernhard Schmid. "Collective Intentionality." In Edward N. Zalta, ed., *The Stanford Encyclopedia of Philosophy*, Summer 2013 Edition. https://plato.stanford.edu/archives/sum2013/entries/collective-intentionality/.
Scott, John. *What Is Social Network Analysis?* London: Bloomsbury Academic, 2012.
Searle, John R. *Making the Social World: The Structure of Human Civilization*. New York: Oxford University Press, 2010.
Searle, John R. *The Construction of Social Reality*. New York: The Free Press, 1995.
Shennan, Stephen. *Genes, Memes and Human History: Darwinian Archaeology on Cultural Evolution*. New York: Thames & Hudson, 2003.
Shennan, Stephen, ed. *Pattern and Process in Cultural Evolution*. Berkeley: University of California Press, 2009.
Shull, George H. "Mendelian Or Non-Mendelian?" *Science* 54 (1921): 213–16.
Spencer, Herbert. "Progress: Its Law and Cause." In *Humboldt Library of Popular Science Literature*, Vol. 17, 233–85. New York: J. Fitzgerald, 1881.
Spengler, Oswald, trans. Charles Francis Atkinson. *The Decline of the West*. New York: A. A. Knopf, 1926 [1918].
Spier, Fred. *Big History and the Future of Humanity*, 2nd ed. Malden, MA: Wiley, 2015.
Spier, Fred. *The Structure of Big History from the Big Bang Until Today*. Amsterdam, Amsterdam University Press, 1996.
Steward, Julian. *Theory of Culture Change: The Methodology of Multilinear Evolution*. Urbana, IL: University of Illinois Press, 1955.
Stringer, Chris, and Peter Andrews. *The Complete World of Human Evolution*. London: Thames & Hudson, 2005.
Stringer, Chris, and Robin McKie. *African Exodus*. London: Cape, 1996.
Studdert-Kennedy, Michael, Chris Knight, and James R. Hurford, "Introduction: New Approaches to Language Evolution." In Hurford, Studdert-Kennedy, and Knight, eds., *Approaches to the Evolution of Language: Social and Cognitive Bases*. Cambridge: Cambridge University Press, 1998.
Swadesh, Morris. *The Origin and Diversification of Languages*. London: Routledge & Kegan Paul, 1972.

Swanson, G. A. "James Grier Miller's Living Systems Analysis (LSA)." *Systems Research and Behavioral Science* 23 (2006): 263–71.
Székely, Tamás, Allen J. Moore, and Jan Komdeur, eds., Social Behavior: Gene, Ecology and Evolution. Cambridge: Cambridge University Press, 2010.
Tattersall, Ian. *Masters of the Planet: The Search for our Human Origins.* New York: Palgrave Macmillan, 2012.
Thompson, E. P. *The Making of the English Working Class.* New York: Pantheon, 1964.
Thompson, John N. "Four Central Points About Coevolution." *Evolutionary Education Outreach* 3 (2010): 7–13.
Thompson, John N. "The Raw Material for Coevolution." *Oikos* 84 (1999): 5–16.
Tomasello, M., Kruger, A., & Ratner, H. "Cultural Learning." *Behavioral and Brain Sciences* 16, 3 (1993), 495–511.
Tomasello, Michael. *Becoming Human: A Theory of Ontogeny.* Cambridge, MA: Belknap Press of Harvard University Press, 2019.
Tomasello, Michael. "Human Culture in Evolutionary Perspective." In M. Gelfand, ed., *Advances in Culture and Psychology*, 5–51. Oxford: Oxford University Press, 2011.
Tönnies, Ferdinand, trans. and ed. Charles P. Loomis. *Community and Society.* New York: Harper & Row, 1963 [1887].
Toynbee, Arnold J. *A Study of History*, 12 vols. Oxford: Oxford University Press, 1933–61.
Trigger, Bruce G. *Sociocultural Evolution: Calculation and Contingency.* Oxford: Blackwell, 1998.
Tuomela, Raimo. *Social Ontology: Collective Intentionality and Group Agents.* Oxford: Oxford University Press, 2013.
Tuomela, Raimo. *The Philosophy of Social Practices: A Collective Acceptance View.* Cambridge: Cambridge University Press, 2002.
Tuomela, Raimo. *The Philosophy of Sociality: The Shared Point of View.* New York: Oxford University Press, 2007.
Turchin, Peter. *Ultra Society: How 10,000 Years of War Made Humans the Greatest Cooperators on Earth.* Chaplin, CT: Beresta Books, 2015.
Turchin, Peter, Thomas E. Curriec, Harvey Whitehoused, Pieter François, et al. "Quantitative Historical Analysis Uncovers a Single Dimension of Complexity that Structures Global Variation in Human Social Organization." *PNAS* (December 21, 2017), E144–E151.
Turner, Jonathan H. *Human Emotions: A Sociological Theory.* London: Routledge, 2007.
Turner, Jonathan H. *Human Institutions: A Theory of Societal Evolution.* Lanham, MD: Rowman & Littlefield, 2003.
Turner, Jonathan H. *On the Origin of Societies by Natural Selection.* Boulder: Paradigm Publishers, 2008.
Turner, Jonathan H. *On the Origins of Human Emotions: A Sociological Inquiry into the Evolution of Human Affect.* Stanford, CA: Stanford University Press, c2000.
Turner, Jonathan H., and Richard S. Machalek. *The New Evolutionary Sociology: Recent and Revitalized Theoretical and Methodological Approaches.* New York: Routledge, 2018.
Tylor, Edward B. *Anthropology: An Introduction to the Study of Man and Civilization.* London: Macmillan, 1881.
Tylor, Edward B. *Primitive Culture.* London: J. Murray, 1871.
Tylor, Edward B. *Researches into the Early History of Mankind and the Development of Civilization*, 2nd ed. London: J. Murray, 1870.

Tyrrell, Toby. *On Gaia: A Critical Investigation of the Relationship Between Life and Earth*. Princeton: Princeton University Press, 2013.

Udehn, Lars. *Methodological Individualism: Background, History, and Meaning*. London: Routledge, 2001.

Van Zanden, Jan Luiten, Joerg Baten, Marco Mira d'Ercole, Auke Rijpma, Conal Smith, and Marcel Timmer. *How Was Life? Global Well-Being since 1820*. OECD Publishing, 2014.

Von Bertalanffy, Ludwig. *General System Theory: Foundations, Development, Applications*. New York: G. Braziller, 1969.

von Neumann, John, and Oskar Morgenstern. *Theory of Games and Economic Behavior*. Princeton: Princeton University Press, 1944.

von Scheve, Christian, and Mikko Salmela. *Collective Emotions: Perspectives from Psychology, Philosophy, and Sociology*. Oxford: Oxford University Press, 2014.

Wallerstein, Immanuel. *The Modern World-System*, Vol. 1. New York: Academic Press, 1974.

Wallerstein, Immanuel, and Paul Starr, eds. *The University Crisis Reader*. New York, Random House, 1971.

Watson, J. D., and F. H. C. Crick. "Molecular Structure of Nucleic Acids: A Structure for Deoxyribose Nucleic Acid." *Nature* 171 (1953): 737–38.

Weber, Max, Guenther Roth, and Claus Wittich, eds. *Economy and Society: An Outline of Interpretive Sociology*, 3 vols. New York: Bedminster Press, 1968.

Wells, H. G. *The Outline of History*. London: George Newnes, 1920.

White, Leslie. *The Evolution of Culture: The Development of Civilization to the Fall of Rome*. New York: McGraw-Hill, 1959.

Wiesner-Hanks, Merry, ed. *The Cambridge World History*, 10 vols. Cambridge: Cambridge University Press, 2015.

Wignall, Paul B. *The Worst of Times: How Life on Earth Survived Eighty Million Years of Extinctions*. Princeton: Princeton University Press, 2015.

Wilkinson, David. "Central Civilization." *Comparative Civilizations Review* 17 (1987): 31–59.

Williams, George C. *Adaptation and Natural Selection: A Critique of Some Current Evolutionary Thought*. Princeton: Princeton University Press, 1966.

Wilson, David Sloan, and Elliott Sober. "Reintroducing Group Selection into the Human Behavioral Sciences." *Behavioral and Brain Sciences* 17 (1994): 585–654.

Wilson, Edward O. *On Human Nature*. Cambridge, MA: Harvard University Press, 1978.

Wilson, Edward O. *Sociobiology: The New Synthesis*. Cambridge, MA: Belknap Press of Harvard University Press, 1975.

Wilson, Robert A. "Test Cases, Resolvability, and Group Selection: A Critical Examination of the Myxoma Case." *Philosophy of Science* 71 (2004): 380–401.

Wimsatt, William. "Developmental Constraints, Generative Entrenchment, and the Innate-Acquired Distinction." In William Bechtel, ed., *Integrating Scientific Disciplines*, 185–208. Dordrecht, Germany: Kluwer Academic Publishers, 1986.

Wirtén, Eva Hemmungs. *Making Marie Curie: Intellectual Property and Celebrity Culture in an Age of Information*. Chicago: University of Chicago Press, 2015.

Wynne-Edwards, Vero Copner. *Animal Dispersion in Relation to Social Behavior*. London: Oliver & Boyd, 1962.

Zaccarella, Emiliano, and Angela D. Friederici, "Merge in the Human Brain: A Sub-Region Based on Functional Investigation in the Left Pars Opercularis." *Frontiers in Psychology* 6 (2015): 1818.

INDEX

A
Abbott, Patrick, 138, 162
academic freedom, 88, 89
Adolphs, Ralph, 76, 79, 160, 161
Aiello, Leslie C., 141, 169
Anderson, David J., 76, 79, 160, 161
Annales, 106, 125
Aoki, Kenichi, 162
Arnett, Joe, 39
arts and humanities, 3, 18, 83
Avery, Oswald T., 96, 112, 113
Avogadro, Amedeo, 86, 114
Axelrod, Robert, 55–57, 134
Ayala, Francisco Jose, 114, 119

B
Barrett, Lisa Feldman, 76–79, 139, 161
Baumol, William J., 110
Becker, Gary, 134
Becquerel, Henri, 99
behavior, animal, 17, 27
behavior, human, 17, 77, 78, 105, 110, 139
   in groups, 18, 53, 57, 110, 139, 142
   in individuals, 18, 53, 139
   in networks, 18, 53
   of Human System, 13
Benedict, Ruth, 104
beneficiaries (in social evolution), 24, 29, 30, 59, 65, 67, 68, 70, 72, 129, 167

Bentham, Jeremy, 53, 95
Berwick, Robert C., 150, 159, 160
Bickerton, Derek, 149, 150, 162
Big Bang, 2, 153
Binet, Alfred, 102
Binghamton University, 127
biological evolution, 2, 4–8, 11, 23–27, 29, 30, 57, 79, 97–99, 107, 113–117, 119–121, 132, 137, 138, 140, 141, 145, 151, 153, 157, 163, 164
biological sciences, 2, 3, 18, 54, 93, 101, 112, 133, 136, 154
Blau, Peter M., 59, 135
Boas, Franz, 103–106, 121, 139
Bohr, Niels, 100, 103
Boltzmann, Ludwig, 99, 100, 114
borrowing of words, 67
Boyd, Robert, 4, 7, 19, 27, 28, 31, 54–56, 120, 121, 127, 128, 141–147, 162, 166
brain, 2, 6–8, 23, 24, 27, 29, 55, 76, 78, 79, 123, 141, 150, 154, 160, 168
Brooke, John L., 165
Brooks, Alison, 155, 168
Buffon, Georges Louis Leclerc, comte de, 84
Burke, James, 153

C

Campbell, Donald T., 54, 55, 117–122, 124, 127, 128, 130, 144, 146, 150
Cann, Rebecca L., 136, 137
capabilities, 79, 123, 148, 150
capitalism, 18, 48, 73, 74, 86, 91, 96, 107, 111, 127, 129, 131
Carneiro, Robert L., 122, 151
Cavalli-Sforza, Luigi Luca, 69, 116, 127, 128, 130, 141–143
Chase-Dunn, Christopher, 49, 152, 164
Childe, V. Gordon, 121, 122, 129
Chomsky, Noam, 66, 123, 124, 148, 150, 159, 171
Christian, David, 2, 153, 155
cladistics, 115
Club of Rome, 126
Coase, Ronald, 63, 124
coevolution, 4, 5, 7, 10, 11, 13, 33, 39, 114, 115, 138, 145, 147, 150, 163
Coleman, James S., 134
collective intentionality, 23, 24, 29, 58–62, 64, 135, 147, 166, 169–171
colonies, 11, 64, 74, 91, 92, 129
Columbia University, 103, 127
commerce, 11, 12, 70, 71, 85
community, 5, 6, 10, 11, 28–30, 32, 38, 42, 46–48, 62, 64, 65, 67–70, 72, 77, 83, 85, 87, 104, 105, 117, 120, 123, 125, 129, 134, 136, 148, 150, 161, 163, 164, 167–170
Comte, Auguste, 87, 95
cooperation, 7, 55, 56, 58, 75, 115, 134, 144–147, 166, 167, 170
Creanza, Nicole, 28, 161, 162
Crick, Francis H.C., 113
cross-disciplinary analysis. *See* multidisciplinary analysis
Crumley, Carole L., 155
cultural evolution, 4, 5, 7, 8, 13, 19, 23, 24, 27, 28, 31, 32, 55, 79, 110, 117–121, 127, 128, 130, 133, 134, 137, 138, 141–147, 150, 154, 159, 161–163, 166, 167, 170, 171
culture, 2, 12, 27, 31, 70, 104, 121, 142–147, 153, 162
   definitions, 31
   group-level, 32, 66, 121, 128, 145, 161
   individual-level, 32, 142, 145

Curie, Marie, 99, 100, 114
Curie, Pierre, 100

D

Dalton, John, 86, 100, 114
Darwin, Charles, 2, 3, 23–25, 53, 57, 87, 91–98, 102, 107, 112, 114, 121
databases (or datasets), 3, 11, 12, 49–52, 111, 121, 126, 127, 129, 153, 165
Dawkins, Richard, 117, 138, 147
decolonization, 73, 109, 110, 125, 130–132
Degler, Carl N., 78, 104
Delbrück, Max, 103, 112, 114
Denisovans, 26, 158
descent with modification, 23, 25, 97
de Zwart, Pim, 165
diaspora, 69, 168
Dillehay, Tom D., 36, 37
Dilthey, Wilhelm, 98
disciplines
   anthropology, 18, 130, 155, 169–171
   archaeology, 69, 169
   biology, 2, 3, 6, 38, 54, 75, 83, 89, 140, 169, 171
   boundaries, 89
   chemistry, 89, 99
   ecology, 27, 38, 69, 98, 155
   economics, 18, 41, 54, 63, 69, 92, 110, 130, 133
   emotions, 75, 170
   environmental science, 38
   genetics, 6, 106, 139
   geology (Earth Sciences), 1, 34, 86, 89, 92
   health sciences, 49
   history, 1, 2, 18, 106, 125, 130, 155
   history of science, 3
   linguistics, 18, 49, 63, 67, 69, 129, 155
   paleontology, 6, 169
   physics, 33, 34, 89, 93
   politics, 18, 63, 106, 125
   psychology, 18, 55, 75, 129, 134
   sociology, 18, 54, 69, 106, 130, 155
   systems analysis, 109, 133, 154, 155
   See also arts and humanities; biological sciences; information sciences; physical sciences; social sciences

discourse on human evolution, 3, 90, 98
diversity, 4, 103, 108, 128, 137, 167
  biological (genomic), 24, 154, 158
  social, 24, 128, 168
DNA
  ancient, 8, 25, 26, 157, 158, 169
  code, 113, 132
  methylation, 6, 116, 128
  mitochondrial, 26, 137, 154
  sequencing, 25, 157
  Y-chromosome, 26, 137, 154
Dobzhansky, Theodosius, 103, 107, 108, 114, 119
dual heritage, 7, 8, 27, 55, 147, 170
Dunbar, Robin I.M., 65, 141, 169
Durkheim, Émile, 105, 107
dynamics
  agriculture, 30, 65
  capitalism, 48, 73
  chaos theory, 133
  epigenetics, 25
  groups, 61, 70, 79
  individuals, 55, 70, 79, 150
  institutions, 1, 30, 61, 64–66, 68, 79, 167, 168
  literacy, 72
  migration, 68
  networks, 61, 70
  speech, 67
  systems, 33, 48, 73, 111, 133

E
Earth System, 43, 112, 136, 165
ecology, 27, 38, 39, 69, 99, 109, 115, 118, 127, 129, 131, 152, 155, 163
economy, 11, 91, 106, 111, 121, 131, 152
Ehret, Christopher, 67, 68, 148, 149
Ehrlich, Paul, 4, 115
Eimer, Theodor, 102
Einstein, Albert, 100
Elias, Norbert, 58, 110, 111
emotions, 17, 24, 28, 46, 61, 63, 75–79, 94, 95, 139, 140, 151, 160, 161, 164, 170
empires, 12, 47, 49, 64, 91, 92, 95, 99, 107, 132

Akkadian, 49
capitalist, 73, 129
Engels, Frederich, 95, 96
environment, 2–4, 7–9, 12, 33, 46, 47, 63, 110, 112, 133, 138, 151
  environmental change, 5, 8, 10, 13, 126, 161
epigenetics, 6, 25, 31, 116, 128, 131–133, 137, 154, 163
ethnicity, 92, 95, 143, 158
eugenics, 98, 99, 102, 103, 108
eukaryotes, 113, 114
European Union (EU), 131, 157
evo-devo, 116, 131, 137, 139, 147, 159
evolution. *See* biological evolution; cultural evolution; institutional evolution; social evolution
exchange
  in networks, 49, 65, 70, 152
  of emotions, 77
  of goods and services, 49, 65, 152. *See also* commerce
  of knowledge, 18, 70, 84, 85, 172
extinction, 140

F
Faraday, Michael, 86
feelings, 17, 76, 77, 89, 95, 160
Feldman, Marcus, 28, 116, 127, 128, 141–143, 161
Fernand Braudel Center, 127
Fichte, Johann Gottlieb, 88
Fisher, Ronald A., 103, 114, 115
Fitch, W. Tecumseh, 148, 150, 159
fitness
  in biological evolution, 23, 24, 29, 30, 115, 137
  in cultural evolution, 24, 27, 32, 138, 145, 147, 162
  "inclusive fitness", 27, 55, 115, 137, 138, 140, 141, 147, 162
  in social evolution, 24, 29, 32, 68, 138, 167
Flannery, Kent, 71, 72, 122, 164
Forrester, Jay, 111, 126
Frank, Andre Gunder, 127, 152
Franklin, Benjamin, 84
Franklin, Rosalind, 113

Friedman, Milton, 124, 132
Friedrich Wilhelm III, King of Prussia, 87
Friedrich Wilhelm University of Berlin, 87, 88, 96

G

Gaia, 43, 44, 112, 136
Galton, Francis, 102
games, 56–58, 111, 134, 142
game theory, 8, 28, 29, 56, 110, 115, 128, 132–135, 140, 154, 166, 167
Gaur, Albertine, 72, 73
gender relations, 12, 13
Geological Society, 87
Gills, Barry K., 152
Glacial Maximum, 37, 135, 164, 170
globalization, 131, 132, 165
Gobineau, Arthur, comte de, 53, 54, 92, 95
Goethe, Johann Wolfgang von, 84
Gogol, Nikolai, 96
Goodrich, Samuel G., 96
Gould, Stephen Jay, 25, 103, 114, 116, 138, 140
Greenberg, Joseph H., 66, 67, 123, 124, 129, 143, 148, 149
groups
  family groups, 10, 45
  groups in networks, 18, 53, 61
  I-groups, 58–62, 65, 135
  interactions of groups and individuals, 59, 60, 77, 94
  language, 10, 11, 47, 129, 135, 141, 148, 160, 167
  we-groups, 58, 59, 61, 62, 64–66, 135
  See also collective intentionality

H

habitat, 68, 69
Haeckel, Ernst, 94
Halbwachs, Maurice, 125
Hall, Thomas D., 49, 152
Hamilton, W.D., 115, 134, 138, 144
Hancock, David, 70, 71
Harari, Yuval Noah, 165

Hayek, Friedrich A., 55, 124, 132
Hegel, G.W.F., 88
Hennig, Willi, 115
hierarchy, 46, 47, 61, 62, 66, 71, 72, 91, 95, 107, 132, 133, 145, 160, 164, 170, 171
  colonial, 91, 99
  gender, 13
  political, 122, 152
  racial, 9, 91, 92, 99, 104, 109, 130
Hobbes, Thomas, 53, 55
Holmes, Arthur, 34, 35, 100, 101
Holocene, 1, 24, 26, 28, 64, 71, 72, 109, 122, 145, 146, 151, 164, 165
Homo
  erectus, 1, 6, 8
  habilis, 6, 117
  neanderthalensis, 2
  sapiens, 1, 24, 28, 118, 141, 148, 149, 154, 163, 169
  See also Denisovans
Hopkins, Terence K., 127
Hornborg, Alf, 155
human evolution, 1, 3–6, 17, 18, 26, 32, 53, 66, 75, 79, 83, 90, 93, 120, 121, 127, 130, 141, 143, 153, 155, 164, 168, 169
human nature, 75, 77–79, 104, 110, 124, 137, 139
Human Relations Area Files (HRAF), 50, 121
human society, 2, 18, 46, 52, 92, 107, 124, 126, 132, 136, 151
Human System, 5, 12, 13, 23, 47, 48, 157, 170–172
Humboldt, Alexander von, 85, 87, 88
Humboldt, Wilhelm von, 87, 88
hypotheses, 79, 120, 121, 142, 150, 159, 162

I

ideology, 79, 84, 91, 99, 119, 124, 131, 132, 137
information sciences, 3, 18, 38, 41, 49, 128
Ingold, Tim, 151, 164

inheritance, 8, 24, 25, 27, 29, 55, 94, 102, 103, 113, 127, 128, 137, 138, 142, 143, 146, 167
Institute for Research on World-Systems (IROWS), 152, 164
institutional design, 58
institutional evolution, 10, 11, 13, 18, 23, 29, 30, 48, 63, 64, 73, 74, 166–172. *See also* social evolution
institutions, social
   agriculture, 30, 64
   business firms, 73
   capitalism, institutions of, 73
   commerce, institutions of, 70
   community, 10, 29, 167
   definitions, 64, 66
   governments, 63, 71, 91
   language, 10, 11, 64
   leadership, 71, 72
   list of social institutions, 64
   literacy, 72
   markets, 70, 124
   migration, 29, 64, 65, 68–70
   schools, 59
   United Nations, 73
   universities, 84, 91
Intergovernmental Panel on Climate Change (IPCC), 112, 131, 136, 154, 157
International Association of Academies, 85, 92
International Biological Program (IBP), 118, 129
International Council of Science (ICSU), 111
International Geophysical Year (IGY), 111, 118, 129

J
Jablonski, Nina, 9, 10
Johannsen, Wilhelm, 102
Johns Hopkins University, 89

K
Kamin, Leon, 54, 138
Kant, Immanuel, 84
Kettlewell, Henry Bernard Davis, 8

Keynes, John Maynard, 105, 124
knowledge, 1, 2, 4, 13, 33, 63, 64, 79, 84–87, 91, 114, 125, 128–130, 136, 140–142, 169, 172
   academic, 1, 83, 84, 88, 91, 109
   disciplinary, 2, 3, 83, 84, 89, 91, 110, 130
   general, 30, 59, 136
   in cultural evolution, 7, 27
Koch, Robert, 101, 112
Kolodny, Oren, 28, 161

L
Laland, Kevin N., 142, 162
Lamarck, Jean-Baptiste, 87
language
   and linguistics, 67, 122, 123, 129, 148, 149, 159, 168, 170
   as social institution, 10, 64, 167
   classification, 123, 124, 129, 143, 148
   emergence, 167, 168
   families, 10, 67, 123
   learning, 11, 119, 123, 167
   migration, 64, 68, 148, 168
   spoken, 10, 66, 149, 159, 160, 162, 167, 168
   syntactic, 66, 67, 135, 148, 149, 162
   universals, 66, 67, 123, 124, 129, 148, 159
   See also borrowing of words; literacy
Lavoisier, Antoine, 85
leadership, 62, 71, 72, 86, 87, 97, 122, 126, 133
Leakey, Louis S.B., 117, 129
Leakey, Mary, 117, 129
Lederberg, Joshua, 113
Lerro, Bruce, 164
Lewens, Tim, 162, 163
Lewin, Kurt, 110
Lewontin, Richard C., 54, 55, 124, 138, 139
Libby, Willard, 36, 111
Linguistic Society of Paris, 122
literacy, 1–3, 18, 64, 72, 73
Locke, John, 53, 55, 78, 139
Lopreato, Joseph, 77, 78, 139, 161
Lotka, Alfred J., 41, 106

Lovelock, James, 43, 44, 112, 136
Luria, Salvador, 112
Lyell, Charles, 86, 87

M
Machalek, Richard S., 165
Malinowski, Bronislaw, 104, 107
Manning, Patrick, 4, 10–12, 15, 17, 18, 29, 30, 32, 46, 48, 64, 67–70, 73, 84, 85, 111, 112, 118, 126, 148, 149, 153, 165, 167
Marcus, Joyce, 71, 72, 164
Margulis, Lynn, 43, 44, 112–114
Marshall, Alfred, 96
Marx, Karl, 96, 130
Maxwell, James Clerk, 93, 99, 100
Maynard Smith, John, 45, 55, 110, 114, 115, 121, 134, 137, 141, 149, 150
McBrearty, Sally, 155, 168
McElreath, Richard, 19, 27, 28, 55, 56, 145, 146, 163
McLean, Paul, 60, 61
McNeill, J.R., 153, 155
McNeill, William H., 125, 126, 130, 153, 159
Mendel, Gregor, 102, 112
Mendeleev, Dmitri, 93
Merge, 150, 160
Mesoudi, Alex, 162
methodological individualism, 55, 105
migration
　as institution, 29, 32, 65, 68–70, 168
　dynamics, 68
　maritime, 168
　routes, 154
　theories, 69, 70
Milankovitch, Milutin, 101, 112
Milanovic, Branko, 74
Mill, John Stuart, 53, 55, 92
Miller, James G., 44–47, 110, 117, 122, 130, 169
Mithen, Steven, 164
molecular clock, 26, 140
Morange, Michel, 103, 139
Morgan, Lewis Henry, 28, 92, 95, 121
Morgan, Thomas Hunt, 102
Morgenthau, Hans, 125

Morris, Ian, 164, 165
Morse, Samuel, 86
Müller, Johannes Peter, 88
multidisciplinary analysis, 4, 27, 165
Murdock, George Peter, 121, 122, 129

N
Nash, John, 110
Neanderthal, 2, 26, 118, 129, 162
networks, 42, 60, 61, 71, 73, 78, 152, 170, 171
　I-group network, 60, 61, 65
　networks as institutions, 61, 65, 66, 70, 84, 170–172
　we-group network, 61, 62, 65, 66
Newton, Isaac, 84, 86, 93, 100
Nirenberg, Marshall, 113
Nostratic, 148, 149
Nowak, Martin, 138, 162

O
Olson, Mancur, 54, 124
ontogenic change, 25
ontology, 12, 15, 17, 49–52, 61, 70, 71
Ornstein, Robert, 153

P
Pääbo, Svante, 26, 157
Palsson, Gisli, 164
parameters, 10, 26, 28, 35, 37, 38, 68, 86
Pareto, Vilfredo, 105, 106, 124
Parsons, Talcott, 110, 111, 134
Pasteur, Louis, 101
Pearson, Karl, 98, 102
Peccei, Aurelia, 126
physical sciences, 3, 18, 33, 89, 93, 99, 111, 129, 132, 135
Piketty, Thomas, 11, 12
Pinker, Stephen, 78, 139, 149, 159, 164
Planck, Max, 100, 163
Pleistocene, 1, 24, 28, 64, 109, 129, 142, 145, 146, 148, 164
Popper, Karl R., 55
population analysis, 92

progress, 23, 24, 90, 95–97, 104, 107, 108, 128, 164
PubMed, 50

R
racial categorization, 9, 109
Radcliffe-Brown, Alfred R., 107
radioactivity, 34, 98–100
Ranke, Leopold von, 88, 96, 97
rational choice, 54, 55, 124, 133, 134, 168
Raven, Peter H., 115
Reade, Winwood, 97
Reagan, Ronald, 133
reductionism, 54, 78, 103, 114, 137–139, 141, 142
regulation, 6, 29, 42, 48, 67, 70–74, 128, 130–134, 154
Reich, David, 26, 157–159
relativity, theory of, 98, 100
religion, 2, 64, 88, 92, 105, 126, 171
representation, 2, 14, 32, 51, 60, 85, 149, 169
  defined, 66
reproduction, 24, 39, 45, 47, 68, 102, 107, 113, 128
  biological, 7, 25, 27, 45, 117
  social or institutional, 29, 45, 64, 66, 73, 167
Richerson, Peter J., 4, 7, 27, 28, 31, 55, 121, 127, 128, 141–147, 162, 163, 166
Rockefeller University, 112
Röntgen, Wilhelm, 99, 100
Rose, Steven, 54, 138
Rousseau, Jean-Jacques, 53
Ruddiman, William F., 37, 43, 135
Ruhlen, Merritt, 149
Rutherford, Ernest, 88, 100, 114

S
Sanderson, Stephen, 98, 151, 152
Santa Fe Institute, 133, 148
scale, 3, 14, 19, 23, 24, 27, 29, 32–35, 37, 41, 44–48, 51, 58, 62–64, 67, 78–80, 84, 89, 91, 92, 95, 96, 98, 105–107, 109, 117, 126, 130–137, 141, 143, 145, 152, 153, 155, 157, 161, 165, 169, 171, 172
schooling, 64, 88
Schumpeter, Joseph, 55, 105
Searle, John R., 57, 64, 135
selection, 3, 4, 6, 7, 9, 24, 25, 27–31, 53, 57, 63, 67, 93–95, 97, 102–104, 110, 112, 117, 119, 120, 124, 128, 138, 141, 143–147, 151, 152, 159, 168
Service, Elman, 122, 124, 142
Seshat, 52
Shennan, Stephen, 8, 144, 146
Shryock, Andrew, 17, 115, 133, 164
slavery, 64, 72, 121
Smail, Daniel Lord, 19, 115, 133, 164
Smith, Adam, 105
Smithsonian Institution, 110
Snooks, Graeme, 151
snowball Earth, 135
Sobert, Elliott, 57
social evolution, 3–5, 13, 18, 19, 23, 24, 28–32, 48, 55, 63, 83, 90, 99, 104, 109, 117–122, 128, 129, 132, 138, 139, 142, 144, 146–152, 154, 161–163, 167–169. *See also* institutional evolution
social learning, 4, 7, 24, 27, 55, 133, 140–147, 153, 154, 168
social movements, 109, 127, 131
social sciences, 2, 3, 41, 49, 54, 55, 63, 83, 89, 92, 93, 95, 96, 98, 103, 107, 110, 114, 121, 128–130, 132–134, 139, 140, 172
Spearman, Charles, 98, 102
Spencer, Herbert, 23, 95, 97, 104, 121
Spengler, Oswald, 106, 107
Spier, Fred, 153–155, 164
Starotsin, Sergei, 149
state, 42, 43, 75, 88, 89, 95–97, 131, 152, 160
Stern, William, 100
Stoneking, Mark, 135
Swadesh, Morris, 121
Stern, William, 102
Stoneking, Mark, 137
Swadesh, Morris, 123
systems

analysis, 38, 41, 42, 44–46, 49, 109–111, 127–129, 131, 133, 136, 152, 154, 155, 169
climatic, 37, 111, 154, 165
closed, 42
ecosystems, 38, 39
general, 18, 41–43, 110, 164
living, 33, 36, 42, 44, 45, 47, 48, 107, 117, 128, 150, 169
migratory, 68
open, 42, 43, 47
social, 4, 12, 18, 42, 45, 57, 73, 117, 119, 124, 130, 152
solar, 33, 41
subsystems, 12, 41–49, 117, 169
See also capitalism; Earth System; Human System; World-System
Szathmáry, Eörs, 45, 141, 150

T

Tattersall, Ian, 6, 7, 118, 158, 162
technology, 24, 26, 38, 69, 122, 129, 131, 132, 152, 163, 168, 169
temperature, 9, 35, 38, 42–44, 86, 93, 99, 100, 112, 136, 157
Thatcher, Margaret, 133
theories
　biological, 25, 30, 41, 54, 96, 133, 140
　cultural evolution, 7, 8, 27, 28, 120, 144, 146, 161, 163
　ecological, 112
　physical, 35, 41, 98
　social, 28, 29, 61, 63, 64, 73, 99, 125, 134, 141, 154, 167
thermodynamics, 34, 36, 42, 87, 89, 93, 99, 153
Thompson, Edward P., 125
Thompson, William, Lord Kelvin, 93
Tishkoff, Sarah, 158
Tomasello, Michael, 146, 147, 159, 162
Tönnies, Ferdinand, 105, 107
tools, 5, 8, 24, 28, 118
　analytical, 79
　stone, 7, 28, 117
Toynbee, Arnold J., 106, 107
trade unions, 91, 124

Trigger, Bruce G., 104, 122, 142, 151
Tuomela, Raimo, 29, 54, 57–60, 64, 65, 135, 166, 167, 169
Turchin, Peter, 28, 32, 52, 121, 163, 164
Turkana, 6
Turner, Jonathan H., 63, 151, 161, 164, 165
Tylor, Edward B., 95, 97, 121

U

Udehn, Lars, 55, 63, 64, 98, 106, 134
ultrasociety/ultrasociality, 28, 134, 144, 146, 166, 170
United Nations, 11, 13, 109, 111, 118, 129, 136, 154
unit of evolution, 25, 27, 29, 57
universities, 2, 3, 62, 83–85, 87–89, 91, 96, 126, 127, 131, 137, 157

V

van Zanden, Jan Luiten, 165
variation, 24, 25, 27, 29, 37, 52–54, 59, 62, 67, 93, 94, 101, 112, 116, 119, 143, 145, 146, 154, 158, 167
Vico, Giambattista, 84
violence, 75, 163, 164
visual art, 66, 118
von Bertalanffy, Ludwig, 41–43, 109, 110
von Neumann, John, 110

W

Wallace, Alfred Russel, 93, 94
Wallerstein, Immanuel, 48, 111, 121, 127, 152
Walras, Léon, 96, 105
warfare, 65, 73
Watson, James, 113
Weber, Max, 55, 105–107, 110, 130
Wegener, Alfred, 100, 111
Wells, H.G., 106
White, Leslie, 28, 121, 122, 129, 142, 150, 165
Wiener, Norbert, 109

Willerslev, Eske, 26, 157
Williams, George C., 54, 57, 106, 120, 124, 145, 167
Wilson, Allan C., 137
Wilson, David S., 57, 144
Wilson, E.O., 53, 78, 116, 117, 137–139, 141, 142, 161–163
World-System, 48, 49, 111, 127, 130, 152, 155, 164, 165, 171. *See also* systems

Wright, Sewall, 41, 103
Wynne-Edwards, Vero Copner, 54, 57, 120, 167

Y
Y-chromosome, 26, 137, 154

The manufacturer's authorised representative in the EU is Springer Nature Customer Service Centre GmbH, Europaplatz 3, 69115 Heidelberg, Germany. If you have any concerns regarding our products, please contact ProductSafety@springernature.com

Printed and bound by CPI Group (UK) Ltd, Croydon, CR0 4YY

23/03/2026

02076664-0003